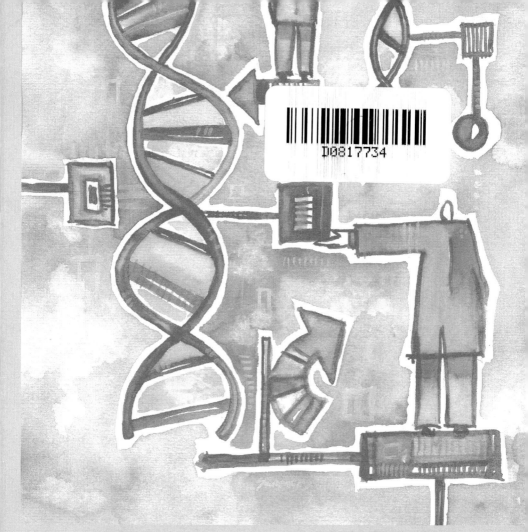

Understanding
BIOTECHNOLOGY

Aluízio Borém • Fabrício R. Santos • David E. Bowen

PRENTICE
HALL
PTR

PRENTICE HALL
Professional Technical Reference
Upper Saddle River, NJ 07458
www.phptr.com

Library of Congress Cataloging-in-Publication Data

Borém, Aluízio.
   Understanding biotechnology/Aluízio Borém, Fabrício R. Santos, David E. Bowen.
      p. cm.
   Includes bibliographical references and index.
   ISBN 0-13-101011-5
      1. Biotechnology—Popular works. I. Santos, Fabrício R. II. Bowen, David E. III.
Title.

TP248.215 .B67 2003
660.6—dc21

                                                                        2002032375

Editorial/Production Supervision: Back Cover Illustration: *DOE Human*
   Jane Bonnell   *Genome Program*
Interior Design: *Gail Cocker-Bogusz*   Manufacturing Buyer: *Maura Zaldivar*
Cover Design Director: *Jerry Votta*   Publisher: *Bernard M. Goodwin*
Cover Design: *Nina Scuderi*   Editorial Assistant: *Michelle Vincenti*
Front Cover Illustration: *Sérgio Ramos*   Marketing Manager: *Dan DePasquale*

© 2003 Pearson Education, Inc.
Publishing as Prentice Hall Professional Technical Reference
Upper Saddle River, NJ 07458

Prentice Hall books are widely used by corporations and government agencies for training, marketing, and resale.

For information regarding corporate and government bulk discounts please contact:
Corporate and Government Sales (800) 382-3419 or corpsales@pearsontechgroup.com

Chapter-opening image for Chapter 3, p. 27, courtesy of Bio-Rad Laboratories, Inc. Chapter-opening image for Chapter 7, p. 87, courtesy of NHGRI. Chapter-opening image for Chapter 10, p. 119, © 1998 The Learning Company and its licensors. Chapter-opening image for Chapter 12, p. 139, from National Aeronautics and Space Administration.

Company and product names mentioned herein are the trademarks or registered trademarks of their respective owners.

Printed in the United States of America

10 9 8 7 6 5 4 3 2

ISBN 0-13-101011-5

Pearson Education LTD.
Pearson Education Australia PTY, Limited
Pearson Education Singapore, Pte. Ltd.
Pearson Education North Asia Ltd.
Pearson Education Canada, Ltd.
Pearson Educación de Mexico, S.A. de C.V.
Pearson Education—Japan
Pearson Education Malaysia, Pte. Ltd.

To Jesus Christ

A. Borém

To my parents, João and Maria,
and my siblings:
Flávio
Fausto
Fabiano
Fabíola

F. Santos

To my friends, and most importantly
Sherrie, Cole, and
my great family

D. Bowen

About Prentice Hall Professional Technical Reference

With origins reaching back to the industry's first computer science publishing program in the 1960s, Prentice Hall Professional Technical Reference (PH PTR) has developed into the leading provider of technical books in the world today. Formally launched as its own imprint in 1986, our editors now publish over 200 books annually, authored by leaders in the fields of computing, engineering, and business.

Our roots are firmly planted in the soil that gave rise to the technological revolution. Our bookshelf contains many of the industry's computing and engineering classics: Kernighan and Ritchie's *C Programming Language,* Nemeth's *UNIX System Administration Handbook,* Horstmann's *Core Java,* and Johnson's *High-Speed Digital Design.*

PH PTR acknowledges its auspicious beginnings while it looks to the future for inspiration. We continue to evolve and break new ground in publishing by providing today's professionals with tomorrow's solutions.

PRENTICE
HALL
PTR

Contents

Foreword

How does society make decisions about new and controversial technologies, like genetic engineering? Who should make those decisions? How should the public be informed about the benefits and risks? These questions barely existed a generation ago. Now, they are central to any discussion about innovation, especially biotechnology and genetic engineering.

No matter how you feel about biotechnology, understanding its basic principles, fantastic potential, and very real risks is essential if you want to have a positive impact on today's public debate. Of course biotechnology is an ancient science, first practiced by people who selected which plants to save, which livestock to cross, which fermented juices to drink. But the fact that biotechnology has been around for a very long time misses the point. Traditional biotechnology reflected the limited impact that people could have on themselves and their environment (ignoring, for the moment, the profound impact of agriculture itself). Modern biotechnology, primarily those technologies made possible by genetic engineering and gene cloning, is both quantitatively and qualitatively different. Given the public's recent experience with the interface between biology and public policy—the Green Revolution, the "War" on cancer, even the government's response to Mad Cow Disease—it is hardly surprising that there is great excitement, great caution, and also great ignorance about biotechnology and the ways it may help or hurt people and the environment.

Some areas of concern about biotechnology are real, no matter what its defenders may say. Ethical conflicts over animal (not to mention human) cloning, deciding who should have access to a person's genetic information, the likelihood that genetically modified fish will escape into the environment and alter the genes in native fish populations; these are not fantasies. The accelerating loss of biodiversity in our food crops as more varieties come from a smaller number of sources, the growing role of life science corporations and their passion for intellectual property protection of genes and genetic products, the disruption in the lives of organic farmers whose crops are fertilized by genetically modified pollen they do not want—these are profoundly difficult challenges for society to understand, let alone rectify.

Yet the benefits of biotechnology are even more compelling. To ignore them would be a mistake of unparalleled magnitude. Biotechnology can reduce the levels of hazardous chemicals released into the environment. Farmers already benefit from this, because they now enjoy a working environment that is far less hazardous. Biotechnology can produce food with better nutrition and novel health-promoting compounds. People in less developed countries may soon be eating fruits and vegetables that produce vaccines to some of the world's most intractable human diseases. Indeed, higher yielding crop varieties not only feed more people, they do so on less land, reducing the stress that agriculture puts on limited land and water resources. Biotechnology has already led to the invention of better diagnostic tools in human medicine. Soon, it will yield new cures for genetic diseases, cancer, heart disease, autoimmune syndromes, even Alzheimer's and other degenerative problems.

Developing a public consensus about biotechnology is a responsibility for everyone in society, not just large corporations and anti-technology activists. *Understanding Biotechnology* is an important and timely book that provides the information and insight to enable readers to participate in the biotechnology debate. Some will read this book and go on to become genetic engineers. Others will read this book and go on to become lawyers, business leaders, journalists, or just better informed citizens. No matter what your motivation for learning more about biotechnology, your knowledge about this subject and your commitment to participate in the public dialogue are essential. *Understanding Biotechnology* will give you the tools you need to participate in this important debate.

Nevin Dale Young
University of Minnesota
St. Paul, Minnesota

Preface

Looking back on the last few years, it is difficult to imagine life without biotechnology. Biotechnology is everywhere you look, from food in the grocery store, to routine medical treatments in the hospital. February 12, 2001 became an important date in human history. Beginning on that date, anyone with access to the Internet could look at a new atlas that contains the entire human genome. Biotechnology is transforming everything we know, believe, expect, and practice in modern society.

Did you know that IBM's largest current project, called Blue Gene, is devoted to genome sequencing and involves a machine that can perform 1 quadrillion calculations per second? Or that Sun Microsystems' largest project is deciphering a protein? Or that Monsanto and Pioneer/DuPont sell almost 80 percent of the world's seed corn? Or that 100 whole animals have already been patented? Biotechnology is expanding well beyond the walls of the laboratory. Its influence has reached the stock market and provided jobs for millions worldwide.

Researchers at a Worcester, Massachusetts, biotechnology company, Advanced Cell Technology, succeeded in creating the world's first cloned human embryos. This and other controversial applications of biotechnology have created many questions and few answers. Geneticists, ethicists, and theologians working together need to reach a balanced view on many highly controversial topics ranging from embryonic stem cell research to ownership of genes.

Biotechnology is part of human life, from big cities to small farm communities. This is especially evident in the agricultural industry. The global area of transgenic crops planted in 2001 was 130 million acres, representing 5.5 million farmers worldwide. This is part of a growing trend that has seen the area planted with transgenic crops increase 19 percent from 2000 to 2001. This indicates a clear need for all to become informed about this growing area of science.

Understanding Biotechnology addresses the background and applications of biotechnology in the many aspects of today's society. We include essential details important for a scientific book, while maintaining an understandable level for most readers. As with highly polarized topics such as biotechnology, balanced arguments are difficult to find and evaluate. *Understanding Biotechnology* offers readers an opportunity to understand this revolutionary science and the issues surrounding its role in today's society, allowing you to develop informed opinions about biotechnology.

Aluízio Borém
Fabrício R. Santos
David E. Bowen

1 History: From Biology to Biotechnology

In this chapter...

Biotechnology, by definition, is any technology associated with the manipulation of biological systems. In fact, well before humans fully understood biology, they were already working with biotechnology in the production of wine and bread. People manipulated the innate properties of microorganisms, plants, and animals to produce goods for their use. With the accumulation of knowledge and increased experience with modern biological techniques, this definition has expanded to include several applications from recombinant deoxyribonucleic acid (DNA) technology to tissue culture, in the production of products and services. What distinguishes the procedures of modern biotechnology are not the principles involved, but the techniques used. For instance, conventional genetic improvements and molecular techniques share several common aspects, such as their objectives. Both approaches aim to develop biological products that are more beneficial to humans. Molecular improvement offers more predictable results than conventional genetic improvement in crop variety. Most desirable attributes in biological products have to be manipulated for improved use by the producer. Conventional improvement for a specific trait is constrained by time and, more importantly, by the existence of the trait in compatible germplasm. For instance, scientists interested in improving disease resistance in crop plants often encounter difficulties finding adequate resistance in available germplasm. At times the desired trait is virtually nonexistent in the species gene pool. With genetic engineering, or molecular improvement, it is possible to transfer a specific gene from a donor to the recipient in a more controlled manner. In genetic engineering, the donor does not have to be sexually compatible. This expands the possible gene pool outside the normal species to include a virtually limitless amount of genes and traits available for improvement.

Human beings, plants, and all living organisms are made up of molecules that contain carbon, hydrogen, oxygen, nitrogen, phosphorus, sulfur, and other elements in minor proportions. All living organisms are made of proteins, which execute most of the cellular functions and are also responsible for fundamental metabolic pathways. These pathways generate all the secondary organic metabolites, such as carbohydrates and lipids, which are essential components of animal and plant tissue.

Biotechnology operates at molecular level, where many of the biological barriers established from speciation disappear. This is possible because all living cells possess DNA, the fundamental molecule of life, which carries genetic information using a simple, yet universal genetic code. DNA codes the proteins that drive all basic functions in humans, animals, plants, insects, and microorganisms. The code simply transforms the sequence of the nucleotides in DNA (A, C, G, or T) into sequences of amino acids, which constitute proteins. Each protein is made by the transcription and subsequent translation of a gene within DNA. Genes and noncoding sequences in a molecule of DNA form the chromosome. These genes and sequences form the basis of species, for they are the basic genetic instructions for each organism. Finally, each biological species has its own genome, composed of all the organized genes in the chromosomes. All of this genetic information necessary for life is contained within each cell of an organism.

ADVENT OF BIOTECHNOLOGY . . . .

This new science and its potential applications have excited many, while creating uncertainty and skepticism in others. Two of the features of biotechnology that might have contributed to the fear that some have of this science are the speed with which it has been adopted in many economic sectors in recent years and the unexpected way its applications have reached the market. Consider, for instance, the elapsed time between the invention of some products and their arrival on the market (Table 1-1). Whereas only 50 to 100 years ago a new invention used to take as long as 30 years to be available to the public, more recent inventions are marketed even before the public becomes aware of them. Note the case of television, which was invented in 1907, but didn't arrive on the public market until 1936. However, it only took 11 years to market biotechnological products after the first transgenic plants were developed. The shortened period of time might not be long enough for society to learn and become accustomed to transgenic products.

Society is basically made up of three subgroups: conservatives, who tend to be resistant to innovations; progressives, who are enthu-

Table 1-1
Elapsed Time Between Development of Some Products and Their
Commercialization

Technology	Invention	Beginning of Production	Elapsed Time for Commercialization
Pen	1888	1938	50 years
Television	1907	1936	29 years
Transgenics	1983	1994	11 years

siastic for new technologies; and moderates, who only adopt new technologies gradually. All three segments of society can be clearly distinguished when considering biotechnology.

As biotechnology started to draw attention from scientists and the public in the 1980s, most people felt unsure about the new science. Debate frequently centered on the possibility of biotechnology solving the problems facing human health and farm production. Unfortunately, the method and the speed by which biotechnology began reaching society seemed to threaten many who were already working to solve the challenges of human health and agriculture.

In many forums, people were debating whether biotechnology, an emergent science, would substitute for classical genetics, the science that has contributed to food production worldwide through improved varieties of corn, soybean, and apples, as well as new lines of swine, poultry, and other livestock. The arrival of biotechnology was unwelcome news to those who had already dedicated a great part of their professional lives to developing improved varieties through traditional techniques. The fear that in vitro studies would replace proven field and greenhouse techniques was a valid concern. Additionally, a lack of good marketing in introducing the new science and its products to society contributed to the apprehension toward biotechnology. Fortunately, debates on the threat of molecular breeding substituting for classical breeding are now an anachronism; today, the two sciences are seen as complementary.

BIOTECHNOLOGY PRIOR
TO THE 21ST CENTURY . . . .

Although microorganisms have been used for many years in the production of wine and bread, it was Robert Hooke's discovery of cells in a piece of cork in 1665 that opened the door to many discoveries and innovations in biology. About 10 years later, Anton van Leeuwenhoek built a microscope with an amplification power of 270X, allowing him to see microorganisms for the first time. The microscope opened the window to a new world previously invisible to man. In the mid-19th century, Matthias Schleiden and Theodore Schwann presented the theory that all living organisms possess cells.

New issues emerged with the new knowledge, causing some scientists to question why offspring tend to resemble their parents. It was in the middle of the 19th century that the monk Gregor Mendel, working in Brno of the Czech Republic, unveiled the secrets of heredity. Mendel noticed that traits of the garden pea plant, such as flower color, plant height, and seed shape, were consistently transferred from parents to offspring. An inquisitive man, Mendel made hundreds of crosses with different pea plants over many years. By recording the segregation of traits from thousands of plants, Mendel discovered that the traits segregated into predictable ratios, season after season. This indicated that discrete units of heredity were passed from generation to generation. These discoveries were presented in forums in the mid-1800s and compiled in Mendel's "Experiments in Plant Hybridization," published in 1865. These discoveries were an important key to the emerging sciences of heredity and genetics. Interestingly, Mendel's work was largely ignored during his life. In fact, his research was virtually lost for 30 years, until three other scientists rediscovered Mendel's work. His simple experiments in the gardens of an Austrian monastery have provided the basis for innumerable further advancements in biology. Today, no genetics or biology class is able to adequately treat the historical aspects of biology without mentioning Gregor Mendel (Figure 1-1).

By connecting a few more pieces of the puzzle of this emerging science, scientists in the first half of the 20th century were able to conclude that something inside of cells was responsible for heredity.

Figure 1-1
Gregor Mendel, considered the father of genetics.

After the addition of dyes for visualization of cells under the microscope, some structures that were stained in a distinct way were identified and called chromosomes (Figures 1-2 and 1-3). Researchers eventually learned that these structures carry the genes that code for the diversity of life.

Table 1-2 shows numbers of chromosomes and estimates of genome size and number of genes in different species.

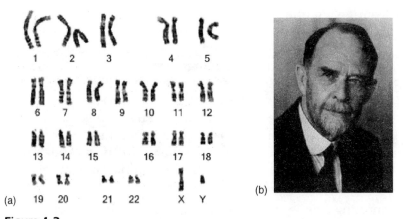

Figure 1-2
(a) Human chromosomes and (b) Thomas H. Morgan, who identified the structures as carriers of genes, the basis of heredity.

Source: (b) © The Nobel Foundation.

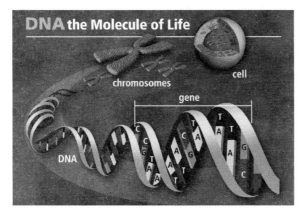

Figure 1-3
DNA, gene, and chromosome, responsible for the likeness between parents and offspring.

Source: Courtesy of DOE Human Genome Program and the web site *http://www.ornl.gov/hgmis.*

In 1941, George Beadle and Edward Tatum (Figure 1-4) established the theory of "one gene, one enzyme," which answered a question that had persisted in the scientific community for many years by explaining how genes are coded instructions for the construction of proteins. Although genes determine human stature, eye color, hair

Table 1-2
Some Genetic Features of Different Species

Species	Number of Chromosomes	Genome Size (Mb)	Gene Number
Human	46	3,000	30,000–40,000
Cattle	30	3,800	35,000
Dog	39	3,000	35,000
Wheat	42	16,000	50,000–75,000
Corn	20	2,500	50,000
Soybean	40	1,100	—
Rice	24	430	25,000
Arabidopis	5	125	26,000

Figure 1-4
Beadle (left) and Tatum, who postulated the theory
of "one gene, one enzyme."
Source: © The Nobel Foundation.

color, and other characteristics, they are not seen, but their products—proteins—are.

In 1944, Oswald Avery identified DNA as the raw material that
contains genes. Starting from that discovery, several research groups
focused their studies on DNA, and the chemical composition was
elucidated quickly. DNA is a molecule made of sugar, phosphate,
and four nitrogenous bases: adenine, cytosine, guanine, and thi-
amine, identified by the initials A, C, G, and T, respectively. Later, scientists realized the four nitrogen bases, also known as nucleotides,
were the alphabet of the genetic code.

The year 1953 marked a landmark for genetics, with the discovery
of the helix-like structure of DNA, by two scientists working at Cambridge University in England: James Watson and Francis Crick, an
American and a British scientist, respectively (Figure 1-5). Their work
revolutionized genetics and accelerated the discoveries of the fine
structure of DNA.

The capability of DNA to code for all the processes of living organisms rests in the order of the genetic alphabet (A, C, G, and T) and is
written in the chromosomes. Genes differ in size (number of letters
or nucleotides) and sequence (order of the nucleotides). For instance,
a gene can have the following sequence: ATGCCGTTAGACTGAAA.
However, the question remained of how the sequence of letters translated into proteins and traits.

Parent
Strands

(a)

(b)

Complementary
New Strand

(c)

Figure 1-5
(a) Complementary strand of DNA and scientists (b) Watson and (c) Crick.

Source: (a) Adapted from *Mapping Our Genes, the Genome Projects How Big, How Fast?*, U.S. Congress, Office of Technology Assessment, OTA-BA-373 (Washington, DC: U.S. Government Printing Office, 1988. (b) and (c) © The Nobel Foundation.

Cracking the genetic code was a puzzle that challenged geneticists: How could only four nucleotides code for the 20 different amino acids that constitute the thousands of proteins found in living organisms?

It was only in 1967 that Marshall Nirenberg and Har Gobind Khorana deciphered the genetic code. They concluded that DNA could be translated by the way the nucleotides (or bases) were organized in groups of three, later called codons (Figure 1-6 and Table 1-3). As there are four nucleotides (A, T, C, and G), there would be 64 different ways to arrange them into different codons. Considering that

Alanine (Ala)	Leucine (Leu)
Arginine (Arg)	Lysine (Lys)
Asparagine (Asn)	Methionine (Met)
Aspartic Acid (Asp)	Phenylalanine (Phe)
Cystine (Cys)	Proline (Pro)
Glutamic Acid (Glu)	Serine (Ser)
Glutamine (Gln)	Threonine (Thr)
Glycine (Gly)	Tryptophan (Trp)
Histidine (His)	Tyrosine (Tyr)
Isoleucine (Ile)	Valine (Val)

Figure 1-6
Nirenberg (above) and Khorana and the amino acids for which the genetic code was deciphered in 1967.
Source: © The Nobel Foundation.

only 20 amino acids exist, the 64 different codons were more than enough to code for each amino acid. Actually, a single codon can represent many amino acids (Table 1-3).

The final products of gene translation, through the genetic code, are proteins. Proteins differ from each other based on the number, type, and order of the amino acids. Proteins can have structural function, such as the collagen and the elastin that are present in the skin, or proteins can carry out metabolic functions, as in the cases of hormones and enzymes.

Until some decades ago, DNA could only be visualized with an imaginative mind; now it can be seen, photographed (Figure 1-7), and even precisely manipulated. As the nature of DNA and genes began to be understood, many questions that existed in the minds of geneticists for many decades were replaced by more applied questions of how this technology could be applied to life. These questions

Table 1-3

Genetic Code for the 20 Amino Acids

First Position	Second Position				Third Position
	T	**C**	**A**	**G**	
T	Phe	Ser	Tyr	Cys	T
	Phe	Ser	Tyr	Cys	C
	Leu	Ser	Stop	Stop	A
	Leu	Ser	Stop	Trp	G
C	Leu	Pro	His	Arg	T
	Leu	Pro	His	Arg	C
	Leu	Pro	Gln	Arg	A
	Leu	Pro	Gln	Arg	G
A	Ile	Thr	Asn	Ser	T
	Ile	Thr	Asn	Ser	C
	Ile	Thr	Lys	Arg	A
	Met	Thr	Lys	Arg	G
G	Val	Ala	Asp	Gly	T
	Val	Ala	Asp	Gly	C
	Val	Ala	Glu	Gly	A
	Val	Ala	Glu	Gly	G

Figure 1-7
Human DNA can now be visualized and photographed.

transformed biotechnology from a science to an important business. In 2001, biotechnology generated revenues of more than $200 billion in the United States alone. Employment and investment opportunities are available as biotechnology releases new products, such as medicines and new crop varieties, and new services like gene therapies and genetic tests.

PRODUCTION OF PROTEINS . . . .

Although DNA stores the genetic information for production of all proteins, it is the ribonucleic acid (RNA) that processes the coded message from the DNA. That is, RNA is more directly used to assemble the proteins by joining amino acids, according to the message coded in the DNA.

Protein production begins with transcription, in which a molecule of messenger RNA (mRNA) is synthesized, copying the message from the DNA. After transcription, mRNA goes to the cytoplasm, where its sequence is translated into the sequence of amino acids that produces the proteins.

The human body possesses more than 100,000 proteins, coded for by about 30,000 genes, according to estimates generated by the Human Genome Project. There are more proteins than genes because many proteins are modified from a single basic protein or are composed of many proteins. Some proteins also arise from different mechanisms of gene expression, a subject too involved to be addressed in this book.

HUMAN GENOME PROJECT . . . .

Scientists are still far from identifying and characterizing all the proteins in the human body. However, incredible strides have been made to provide a foundation for protein research. This reaches to the source of proteins and ultimately the source of life. This foundation is laid by deciphering the entire genome sequence, or DNA (gene) sequence of an organism. Beginning with bacteria, microscopic worms, and yeast, scientists and computational biologists have expanded DNA sequence information to include certain animals and plants. The ultimate goal of DNA sequencing is the human genome. This genome sequence would allow the understanding of the basis of human life by identifying the order of DNA nucleotides. To accomplish this goal, many groups have come together to work on the Human Genome Project.

The sequencing of the human genome, which is finding the order of the more than 3 billion nucleotides (A, T, C, and G) in the human chromosomes, is being accomplished by two independent groups of scientists. The two versions of this sequence were published in the magazines *Nature* and *Science* in February 2001. One group, formerly led by Craig Venter, is Celera Genomics, of Rockville, Maryland, a company started in 1998. The other research group is the result of a consortium of public agencies with laboratories in several countries.

The sequence of the human genome carried out by the public sector, now led by Francis Collins, has a budget of more than $3 billion. The major sponsors were the U.S. Department of Energy and the National Institutes of Health (NIH), as well as the Wellcome Foundation of England. The current map covers about 95 percent of the human genome and has been found to be 99.96 percent accurate. This work has revealed, in a surprising way, that the human genome only has about 30,000 genes instead of 70,000 to 140,000, according to previous estimates. With a DNA sequence of over 3 billion base pairs (bp) and considering the average gene size of 3,000 bp, it is estimated that only 3 percent of the human genome actually codes for some protein. This means that about 97 percent of the human genome has seemingly no coding function; that is, most of the nucleotide sequences in human DNA do not code for genes. This nonfunctional portion of DNA has, for lack of a more accurate term, been called "junk DNA," and its function and purpose have yet to be understood. More important, the data from the Human Genome Project has also revealed that each human being, independent of apparent differences, is about 99.9 percent identical to any other person.

With so much interest and emphasis on the Human Genome Project, what are the practical applications of the sequence of the human genome? The information will help in the early diagnosis of disease, an understanding of the predisposition to genetic diseases, and in genetic counseling, for example. For instance, the sequence of the human genome allows geneticists to understand why certain people have a predisposition to heart disease, and it will eventually lead to the development of new drugs specifically developed to combat the cause of disease and not the symptoms alone. Sequencing the genome will make available basic scientific knowledge for the development of gene therapies for incurable diseases, such as diabetes, muscular dystrophy, cystic fibrosis, Parkinson's disease, and Alzheimer's disease.

By the beginning of 2002, geneticists had already isolated about 13,000 human genes and learned about their functions, including those that code for eye color, circulatory proteins, and genes that when mutated cause a predisposition for developing breast cancer and prostate cancer. All this complex information is contained in each and every cell of the human body. If it were possible to stretch out the incredible amount of information contained in the DNA of all the chromosomes in a single human cell, it would reach about seven feet. If the DNA of all the cells of the human body were stretched out and aligned, it would be enough to cover the distance from Earth to the moon about 8,000 times. Incredible packaging mechanisms allow this information to be stored within each tiny cell.

Concerns of the Sequencing Project

The importance of the Human Genome Project has raised many concerns, both biological and ethical. These questions are being addressed as the information generated by the project is being processed and used by people worldwide.

- Privacy and confidentiality of the genetic information: Who owns the genetic information?
- Right to use the genetic information by insurance companies, employers, courts, schools, adoption agencies, and so on: Who should have access to individual genetic information and how should it be used?
- Psychological impact and stigma attached to an individual's genetic differences: How does personal genetic information affect an individual and society's perception of that individual? How does genomic information affect members of minority communities?

- Reproductive issues, including informed consent for complex and potentially controversial procedures, use of genetic information in reproductive decision making, and reproductive rights: Do health-care personnel properly counsel expectant parents about the risks and limitations of genetic technology? How reliable and useful is fetal genetic testing? What are the larger societal issues raised by new reproductive technologies?

- Clinical issues, including the education of doctors and other health service providers, patients, and the general public in genetic capabilities, scientific limitations, and social risks, including implementation of standards and quality-control measures in testing procedures: How will genetic tests be evaluated and regulated for accuracy, reliability, and utility? (Currently, there is little regulation at the federal level.) How do we prepare health-care professionals for the new information relating to genetics? How do we prepare the public to make informed choices? How do we as a society balance current scientific limitations and social risk with long-term benefits?

- Uncertainties associated with genetic tests for susceptibilities and complex conditions (e.g., heart disease) linked to multiple genes and environmental interactions: Should testing be performed when no treatment is available? Should parents have the right to have children tested for adult-onset diseases? Are genetic tests reliable and interpretable by the medical community?

- Conceptual and philosophical implications regarding human responsibility, free will versus genetic determinism, and concepts of health and disease: Do people's genes make them behave in a particular way? Can people always control their behavior? What is considered acceptable diversity? What is the line between medical treatment and enhancement?

- Health and environmental issues concerning genetically modified (GM) foods and microbes: Are GM foods and other products safe to humans and the environment? How will these technologies affect developing nations' dependence on the West?

- Commercialization of products including property rights (patents, copyrights, and trade secrets) and accessibility of data and materials: Who owns genes and other pieces of DNA? Will the patenting of DNA sequences limit their accessibility and development into useful products?

Incredible advancements have occurred in the last century. Genetics has emerged from the observations of basic biology and questions over the inheritance of traits from parents to offspring. The science has since been enhanced to reveal the sequence of nucleotides in genes that code for proteins. This has helped scientists understand that the genetic code is universal, meaning a similar nucleotide sequence, from soybean, humans, cattle, or bacteria, will result in the production of the same protein with a similar function. This is the basis of biotechnology, allowing genetic engineers to transfer genes among different species, with the objective of transferring desirable traits. With a basic understanding of biology, one can begin to understand biotechnology.

2 Genetic Engineering

Dozens of different kinds of engineers exist. Generally, the most common types are civil, agricultural, mechanical, and electrical engineers. More recently, other types of engineers have appeared, such as nuclear and genetic engineers, among others. Each one of these professionals uses different equipment to accomplish their engineering work. Some engineers use welders, and others seismographs, but all need tools to practice their trade. A genetic engineer uses biological tools to build genes instead of bridges, buildings, and cars. It would be almost impossible to list all the equipment used in a recombinant DNA laboratory, but basic tools are common to all genetic engineers (see Figure 2-1 for an example of DNA equipment). The construction of artificial genes and their transfer into other organisms (a process called transgenics or recombinant DNA) can actually be accomplished at a rather simple laboratory, in comparison with the sophisticated laboratories used for nuclear engineering.

Some pieces of equipment, such as the thermocycler or polymerase chain reaction (PCR) machine, have been used for many purposes and in different situations. The equipment can be used to detect viral diseases before their symptoms are evident, or to distinguish between different soybean varieties, which are virtually identical as far as plant and seed traits are concerned. In this case, visual analysis is not sufficient to differentiate between similar soybean varieties. PCR analysis has also been used to detect residues of transgenic crops in food. In the field of the human genetics, thermocyclers are used for paternity tests, individual identification, and genetic tests. For instance, blood samples found at crime scenes have been used to incriminate some suspects and exonerate others. This process can even

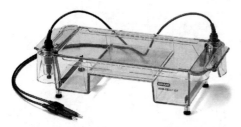

Figure 2-1
Simple piece of equipment used to separate fragments of DNA samples.
Source: Courtesy of Bio-Rad Laboratories, Inc.

be used to identify suspects from the DNA left in residue of saliva used to glue stamps on an envelope. The resolution, power, and potential applications of this technology are enormous, as is discussed in further detail in Chapter 10, "Forensic DNA."

Most of the discoveries in genetic engineering resulted from experiments done with bacteria. These organisms present a series of characteristics that facilitate the studies of genetics. Bacteria multiply quickly, can be preserved easily, and grow in relatively simple and inexpensive tissue culture media. The great genetic diversity found in bacteria is also another factor that favors their use in genetic studies. In nature, bacteria exist in nearly all environments, and some can survive in the most extreme temperature, pressure, and chemical conditions.

Another characteristic of bacteria is the form in which their DNA is organized. In eukaryotes (plants, animals, mushrooms, etc.), DNA is packaged into chromosomes within the nucleus of the cell. Bacteria are prokaryotic organisms, meaning they lack a well-defined nucleus, and the DNA is located in circular chromosomes in the cytoplasm. Many types of bacteria also possess other small circular chromosomes, called plasmids. These plasmids frequently possess genes for antibiotic resistance, which are passed from one bacterium to another, creating an opportunity for gene transfer. Genetic engineers have been manipulating plasmids and transferring genes via plasmids between different species. This is made possible by the ability of the plasmids to be expressed when transferred to another species.

CENTRAL DOGMA . . . .

The genetic material of any organism is the substance that carries the information that determines its life cycle and its characteristics. There is a procedure by which this genetic material is used in living processes; this is the central dogma of genetics. Before the development of modern genetics, it was commonly believed that the substance responsible for heredity was a protein. Once DNA was recognized as the genetic material, the central dogma was established. This states that the information contained in DNA is translated into protein through the processes of transcription and translation. The protein is then used in all life processes, from cell division to electron transport in photosynthesis. For this to occur, DNA is copied (transcribed) into mRNA, and the mRNA is used as template for production of the protein in a process called translation. The message coded by the mRNA sequence gene is translated into a sequence of amino acids, the basic components of protein. Cells cannot produce a protein by simply aligning amino acids; they need to use an RNA template. Additionally, the use of an intermediate mRNA template in protein synthesis reduces the risk of damage to the DNA that can occur from repeated use. Additionally, the central dogma postulates that the intermediate mRNA molecule, a direct copy of DNA, can be used repeatedly in protein synthesis.

Therefore the main points of the central dogma are as follows:

- Genes are made of DNA.
- Genes carry information about structures and biological functions, coded by nucleotides (A, C, G, and T).
- The genetic information is converted in an mRNA molecule.
- The mRNA defines the number, type, and order of amino acids in proteins.
- The protein structure is determined by the linear order of amino acids.
- The three-dimensional protein structure defines its biological function.

GENE STRUCTURE . . . .

DNA is a sequence of nucleotides that code for genes. A gene is the smallest physical and functional unit of heredity, and it codes for a specific biological structure or function. Exons are parts of the DNA within a gene that form the RNA transcript. The outline in Figure 2-2 illustrates a typical gene of eukaryotic organisms: animals and plants, showing the promoter, a DNA sequence preceding a gene that contains regulatory sequences controlling the rate of RNA transcription. Promoters control when and in which cells a given gene will be expressed. To the right of the promoter is the coding sequence with

Figure 2-2
Simplified structure of a typical gene from eukaryotic organisms.

the exons and introns, which are transcribed; in other words, they are copied into an mRNA molecule during the expression of the gene. The introns fill in the spaces between the exons in the gene locus. The introns are transcribed, but are spliced out during mRNA processing, just before translation (production of the protein). The exons contain the coding sequence for protein to be synthesized.

Although molecular genetics has generated much more detailed information about the sequence and function of each part of the gene locus, the understanding that genes possess several other parts besides the coding region is enough for the scope of this book.

VIRUSES . . . .

Viruses are microorganisms in the gray area of what is living and nonliving. Viruses are made of a protein envelope, which surrounds the genetic material (DNA or RNA). Although viruses have their own genetic material like many other living organisms, they do not possess the capacity to reproduce by themselves. For that, they need to use the machinery of living cells to produce a new virus.

The viruses that live in bacteria are called bacteriophage. They inject their DNA into bacterium, leaving their protein envelope outside. Inside of the bacterium, the virus is a filament of nucleic acid that contains coded information for the synthesis of new virus particles, which can be released with the bacteria lyses. The genetic engineering came about from the observation of how viruses use cells of other organisms or bacteria to express their own genes. In that sense, the viruses can be considered genetic engineers. One of the first experiments of genetic engineering was carried out using a bacteriophage as a true Trojan horse, to introduce DNA from other organisms into the bacterium.

One of the requirements for genetic engineering experiments is the production of DNA fragments that contain the desired information. At the beginning of the molecular biology era, DNA used to be cleaved with vibration by ultrasound waves. One of the difficulties that scientists faced in those experiments was the random fashion in which the DNA fragmented. In 1970, however, Dr. W. Arber discovered that bacteria themselves possess a mechanism to specifically cut DNA at certain sequences. Bacteria produce proteins called restriction enzymes that cleave the DNA at specific recognition sites. It was only after the discovery of the restriction enzymes that genetic engineering became a reality. The restriction enzymes were developed as a defense mechanism of bacteria against viruses. Viruses inject DNA

Table 2-1
Partial List of Restriction Enzymes Used in Molecular Biology

Enzyme	Restriction Site	Source
Aat I	G ACGT' C C' TGCA G	*Acetobacter aceti*
Acc III	T' CCGG A A GGCC' T	*Acinetobacter calcoaceticus*
*Bam*H I	G' GATC C C' CTAG' G	*Bacillus amyloliquetaciens*
*Eco*R I	G' AATT C C TTAA' G	*Escherichia coli* RY 13
*Eco*R V	GAT' ATC CTA' TAG	*Escherichia coli* J62 PLG74
Nco I	C' CATG G G GTAC' C	*Nocardia corallina*
Xba I	T' CTAG A A GATC' T	*Xanthomonas badrii*

Note: ' indicates the cut site.

into bacteria and use their bacteria as a mechanism for reproduction. The bacteria, however, develop a mechanism that fragments the exogenous DNA using restriction enzymes. The restriction enzymes recognize the foreign DNA by means of certain specific nucleotide sequences. Different enzymes recognize and cut the DNA at different sites. Using this knowledge, restriction enzymes became essential tools for the genetic engineer to cut DNA into fragments and build new genes. Hundreds of different restriction enzymes exist, many of which are frequently used in biotechnology. Table 2-1 shows some restriction enzymes and their recognition sites for cleavage.

GENE MANIPULATION ·

Before beginning any genetic engineering project, it is necessary to obtain a reasonable amount of relatively pure DNA, which is then cut and ligated to build the new gene. Today, several companies make DNA extraction and purification kits, making the genetic engineering process much simpler. The basic procedure requires the releasing of the DNA from the cells and purification of the DNA to be used in the experiments.

In a typical extraction and purification procedure for plasmid DNA, bacterial cells with the desired plasmid are lysed (broken up) under alkaline conditions and the crude lysate (remains of the cells) is purified using either filters or centrifugation. The lysate is then loaded onto an apparatus where plasmid DNA selectively binds under appropriate low-salt and pH conditions. RNA, proteins, metabolites, and other low-molecular-weight impurities are removed by a medium-salt wash, then plasmid DNA is released in high-salt buffer. The DNA can then be concentrated and desalted for genetic engineering uses.

General Steps

- Grow the bacteria in liquid culture.
- Centrifuge the bacterial suspension to concentrate the bacteria.
- Discard the supernatant (the liquid part that remains above the pellet).

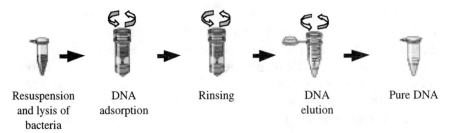

Resuspension   DNA   Rinsing   DNA   Pure DNA
and lysis of   adsorption   elution
bacteria

Figure 2-3
Purification of plasmid DNA after the precipitation of the bacteria.

- Resuspend the bacteria pellet in a solution with RNAse (enzyme that degrades RNA).
- Add a buffer to promote an alkaline lyse of the bacteria.
- Neutralize and adjust the saline conditions of the suspension with the buffer.
- Centrifuge to separate proteins and other impurities.
- Adsorb the plasmid DNA onto a membrane by filtration.
- Rinse the membrane with a solution containing ethanol.
- Elute plasmid DNA from the membrane with EDTA, a chemical substance that preserves the integrity of DNA.

Figure 2-3 displays the latter steps of this procedure.

ENGINEERING GENES . . . .

Once the DNA has been obtained, it is necessary to cut the DNA into pieces to be used for the engineered gene. Restriction enzymes are used for cutting the DNA at specific sites. As seen in Table 2-1, most restriction enzymes cut the DNA into diametric fragments, as opposed to symmetric fragments. That cut leaves the DNA double helix with a small sequence of nonpairing bases that overhang on the end. These regions of DNA are generally used for ligation, or joining with other DNA fragments. DNA fragments cleaved with a single restriction enzyme or with complementary enzymes can be ligated to each other because the overhanging regions are complementary and will bind together. The ligation of fragments is facilitated with addition of the enzyme DNA ligases. The true art of genetic engineering is putting together the parts of puzzle, where each DNA fragment must be placed in right order and orientation so the gene is functional. As scientists know the sequences of genes encoding important traits or proteins, the information is used to engineer genes that can be used in a variety of applications.

Genetic engineers are able to manipulate tiny pieces of DNA with enzymes to create new genes and DNA sequences used in biotechnology. Relatively simple tools in a small laboratory are needed for these engineers to practice their craft. The products that result from these methods can then be used in many applications of biotechnology.

3 Transformation

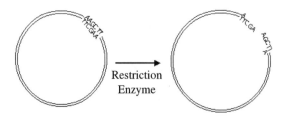

Figure 3-1
DNA cleavage of a plasmid using restriction enzyme.

Genetic transformation in its most basic form is the introduction of transgenes (foreign genes) into an organism in a way that they might be expressed. This technique, also called *genetic engineering*, allows for the transformation of bacteria with genetically engineered plasmids that possess resistance to antibiotics as well as the transformation of a bovine reproductive cell with a transgene for hormone production.

The final objective of genetic engineering, or recombinant DNA technology, is the stable and inheritable expression of a new trait in a different organism or individual. This is done through proper construction of a vector to carry the transgene. Plasmids, retroviruses (RNA virus), and bacteriophages are especially important as vectors (means to deliver the transgene) in the process of transformation. In this process, genetic engineers cut and rearrange DNA fragments to

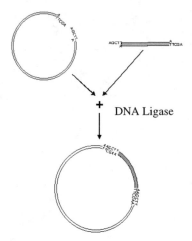

Figure 3-2
Ligation of a transgene with a vector plasmid.

build a genetic construct (transgene), which is finally inserted into a vector (Figures 3-1 and 3-2).

Hebert Boyer and Stanley Cohen achieved the first successful genetic transformation in 1973, when they constructed a gene with portions of DNA from bacteria and an amphibian. These scientists wanted to express antibiotic resistance in an organism that lacked the trait. With the successful use of enzymes and vectors, these men pioneered the use of genetic engineering and transformation. Their work is the basis of much of the current work in biotechnology.

PRINCIPLES OF GENETIC TRANSFORMATION . . . .

The term *genetically modified* is frequently used to describe organisms that were genetically transformed or engineered. The science of genetic engineering was developed with the objective of building genes for genetic transformation. Genetic transformation systems possess three main components:

- A mechanism for introduction of the foreign DNA into the target cell.
- A cell or tissue suitable for transformation.
- A method for the identification and selection of transformed cells or individuals.

Success in transformation for any species depends on these three components. Obviously, each one must be optimized and, therefore, as technology develops, transformation should become a more routine activity. The final objective in transformation is the introduction of a new trait in an individual. When the desired trait exists in any other sexually compatible individual, the first alternative should be to transfer the trait through crossing and selection, as has been done in conventional breeding since the 19th century. Modern soybean, corn, cotton, and wheat varieties, as well as swine, cattle, and poultry lines used in agriculture to feed the world, were initially obtained by traditional methods of crossing and selection.

One of the main limitations of conventional genetic improvement is that the breeder is limited to traits among species that are sexually compatible. For instance, the field bean is a species rich in sulfur-containing amino acids. However, beans are naturally deficient in lysine. On the other hand, rice is naturally rich in lysine, but deficient in sulfur-containing amino acids. It is not possible to naturally cross these species, so the conventional plant breeder is unable to develop a new field bean variety with elevated lysine levels or a rice cultivar rich in sulfur-containing amino acids. Genetic transformation allows the exchange of genes between organisms previously limited by sexual incompatibility. With genetic engineering and transformation, it is possible to transfer genes among bacteria, animals, plants, and viruses. In fact, one of the areas of research in biotechnology is the improvement of nutritional profiles in crops. New, more nutritional bean and rice varieties can now be developed through advances in genetic engineering.

The basic tools for genetic transformation are restriction enzymes, which are used to cut DNA at specific sites, and ligases, which catalyze the joining of DNA fragments, as seen in Chapter 2, "Genetic Engineering." Using the right restriction enzymes, it is possible to cut the circular bacterial plasmid DNA, causing it to linearize. With a ligase, it is possible to add other DNA fragments containing the gene of interest and join them to the linearized plasmid. Under the right conditions, the ends of the plasmid, now with the added DNA fragments, rejoin to create a new circular plasmid with some DNA modifications. The new plasmid can be introduced into certain bacteria through a process called *electroporation*, and the bacteria can then be used to transfer the transgene to the target species. If the plasmid DNA is integrated into the genome of the recipient species and the transferred genes are expressed, the individual is considered to be transformed or transgenic.

METHODS OF GENETIC TRANSFORMATION . . . . .

Among the several methods of plant transformation, four have yielded the best results: *Agrobacterium* species-mediated transformation, microprojectile bombardment, microinjection, and direct transformation. Each of these methods has merits and limitations and is used in specific situations. At this time there is no single technique that is suitable for all species.

Agrobacterium Mediated Transformation

Tumors and uncontrolled cellular growth in plants can occur due to genetic factors or bacterial and viral infections. An example is crown gall in plants, where tumors are caused by bacteria that causes uncontrolled growth on the stem of the infected plants. This problem is caused by *Agrobacterium tumefaciens*, a soil bacterium that infects some plants because of a wound on the plant. Plasmids present in the bacteria are responsible for tumor growth after infection by *A. tumefaciens*. The bacteria are able to recognize wounds on the plant, and this induces the transfer of the bacterial plasmid into the plant. The plasmids are capable of integrating into the DNA of the host plant, causing uncontrolled plant growth and the formation of tumors. The ability of *A. tumefaciens* to efficiently transfer plasmid DNA into the host has made it important in early studies in genetic transformation.

Agrobacterium tumefaciens was the first vector used for introduction of foreign DNA in plant cells. Although *Agrobacterium* has only been used to infect dicot plant species, such as soybean, tomato, pea, and cotton, the protocol has been modified to allow the bacteria to infect some monocot (grass) species as well. Many research groups working with plants have found this to be the preferred transformation approach. Another soil bacteria, *Agrobacterium rhizogenes*, causes the growth of secondary roots after infection. This bacterial species has also been used for plant transformation.

The basis of this transformation method is the bacterial plasmid, which contains the genetic sequence that is integrated into the host genome. One of the most important parts of a plasmid is the region responsible for the translocation of its DNA into the host plant genome. This is called transfer DNA (T-DNA), and this area of DNA is key to the tumor growth in infected plants. The region is located between the right border and left border (RB and LB in Figure 3-3) of the plasmid. Plasmids also contain other important DNA sequences; some of them control the production of auxin and cytokinin, two important plant hormones involved in tumor formation. With the use of the restriction enzymes, a transgene can be introduced between the

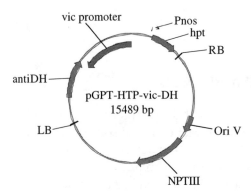

Figure 3-3
Plasmid used in soybean transformation.

right border and left border of the plasmid, allowing the bacteria to transfer novel genes into the recipient plant.

One of the techniques used for transformation mediated by *A. tumefaciens* uses leaf disks (Figure 3-4). Leaf disks of about 6 mm in diameter are cultured on a tissue-culture media containing *A. tumefaciens* with plasmids containing the transgene. After approximately a month of incubation in the tissue culture medium, seedlings start to develop on the leaf disks. Through selection methods, transgenic seedlings are identified for whole plant regeneration.

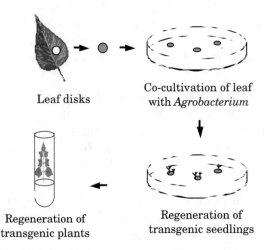

Figure 3-4
Agrobacterium tumefaciens mediated transformation.

Microparticle Bombardment

This technique has also been called *microprojectile acceleration* or *biolistics*, but microparticle bombardment is the formal name for the machine called a gene gun. This method, developed at Cornell University, was designated biolistic (biologic + ballistics = biolistic), because high-speed microscopic projectiles (microprojectiles) are accelerated into the cells to be transformed.

This transformation method consists of the acceleration of a macroprojectile loaded with millions of tungsten or gold microspheres about 1 µm in diameter (microparticle). The microspheres are coated with the transgene, or DNA of the gene of interest. Microspheres have a high specific mass, allowing them to acquire the needed momentum to penetrate the target cells. The macroparticle is propelled in the direction of the cells at high speed, but it is retained, after a small distance, on a steel mesh so that the microparticles continue in the direction of the target cells (Figure 3-5). Helium gas at high pressure is used to propel the macroparticle, and the acceleration chamber operates under a partial vacuum, which allows for improved microsphere movement. Once inside the target cells, the

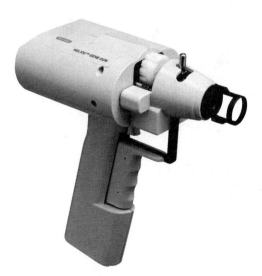

Figure 3-5
Gene gun used in transformation.
Source: Courtesy of Bio-Rad Laboratories, Inc.

DNA coating the microspheres is released and can be integrated into the plant's genome.

Many of the commercial transgenic crop varieties on the market today were developed using the gene gun. However, due to its cost and the complex integration patterns resulting from this method, several research groups are reducing its use.

Microinjection

This method was developed for animal transformation but has also been extended to plants. Although very difficult and laborious, DNA microinjection has yielded positive results and has been used in several laboratories.

In this technique, microcapillary needles are used to introduce DNA directly into cells (Figure 3-6). Each cell to be transformed must be manipulated individually. One of the advantages of this method is

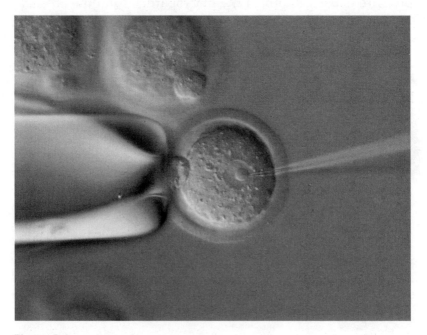

Figure 3-6
Transformation through microinjection.
Source: Courtesy of Manfred Baetscher and Anne Bower, Department of Comparative Medicine, Oregon Health Sciences University.

that the optimum amount of DNA can be injected into the target cells, which helps to ensure optimal integration. Positive results have already been obtained in several crop species such as corn, wheat, soybean, tobacco, and rice, and in animals like salmon, cattle, and swine.

Direct Transformation

Transformation using direct methods was accomplished soon after the first *Agrobacterium*-mediated transformation. These methods use protoplasts (cells after the removal of the cellular wall) as targets for transformation. This is a simple method that consists of adding great amounts of transgenic plasmids to a protoplast culture, which guarantees that a small proportion of the protoplasts will be taken up (assimilated) by the plasmids. The assimilation rate can be increased with the addition of polyethylene glycol (PEG) or the use of an electric discharge (electroporation). No barrier to direct transformation has been detected, indicating that this method can be used with virtually any species. The problem with this method lies in the difficulty of regenerating a whole plant starting from protoplasts. Therefore, it has not been used as widely as the other methods.

BOTTLENECKS IN TRANSFORMATION . . . .

Tissue culture has been identified as one of the largest obstacles in the development of transgenic plant products. It is necessary to develop protocols that allow the regeneration of whole individuals from the transformed cells or tissue. One of the difficulties faced by scientists is that regeneration methodologies work well with some, but not all species or germplasm within a species. This severely limits the spectrum of individuals that can be transformed. In many cases, the procedure has been the transfer of the transgene through classical genetics and breeding methods. An example of this is in the genetic transformation of wheat. Genetic transformation of most wheat varieties is very difficult because of problems in tissue culture. One variety, Bobwhite, is the exception, and protocols have been developed for the transformation of this wheat variety. Once a gene has been successfully transferred into Bobwhite, it can be moved into other varieties through traditional breeding methods.

Another difficulty associated with the use of tissue culture in transformation is somaclonal variations. Plants produced from tissue culture have higher mutation rates and the appearance of abnormal variation. This is due to the delicate environment in which cells are cultured. Many times, the cultured plants have problems associated with the cell cultures and not from the transgene integration.

Transformation methods currently in development promise to revolutionize the introduction of genes in plants. Some of these methods are already being used with the model plant *Arabidopsis thaliana*, commonly known as mouse ear cress. One of the methods involves the submersion of floral buds in a solution containing plasmids bearing the transgenes. Another alternative technique, still in development, is the transformation of seeds mediated by *Agrobacterium tumefaciens*. Although the methods have been used with success in *Arabidopsis*, the literature does not report its use in crop species. The key aspect of these two methods is that transformation is carried out without the need to regenerate plants through tissue culture. These methods are exciting because the transformation procedure works on the seeds that can then be planted to identify transgenic individuals.

GENETIC ENGINEERING PRODUCTS . . . .

Genetic transformation has developed several new products with impacts on society, from medicines to food products with better nutritional quality. The largest commercial success of genetic engineering was the production of human insulin in transgenic bacteria in 1980. Since then, many other products have been released.

The first genetically engineered crop variety to reach the market was the tomato variety Flavr Savr, developed by the Calgene Company, located in Davis, California. This product, introduced to the market on May 21, 1994, was developed with the introduction of two novel genes in a tomato plant. The first gene was a reverse copy of the poligalactonurase gene, which codes for an enzyme that breaks down cellulose. The introduction of this gene in the reverse form, also called *antisense*, resulted in low production of the poligalactonurase enzyme. Consequently, ripe tomato fruits do not lose their firmness because the cell wall of these fruits, which is made of cellulose, does not degrade as rapidly as it does in normal tomatoes. The

Table 3-1
Some of the Genetically Transformed Species

Plants		Animals
Canola	Rice	Cattle
Corn	Soybean	Monkey
Cotton	Sunflower	Mouse
Eucalyptus	Tobacco	Pig
Grape	Tomato	Salmon
Papaya	Wheat	
Potato		

second gene transferred in the development of Flavr Savr codes for resistance to the antibiotic kanamycin. This gene works as a reporter or marker to facilitate the identification of transformed individuals. Table 3-1 includes a partial list of transformed crop species. The introduced traits in those species were, for the most part, tolerance to herbicides, resistance to pests, and nutritional quality.

POTENTIAL OF TRANSFORMATION . . . .

The objectives of variety development through biotechnology are the same as conventional genetic improvement. Most of the desired traits are improved yields, increased vigor, pest resistance, and nutritional quality. However, biotechnology allows the development of varieties with traits that cannot be developed by conventional breeding.

Addition of New Functions

Altered Forms of Enzymes

The introduction of genes coding for a structurally modified enzyme can result in its insensibility to certain chemicals or environmental conditions. For example, the herbicide glyphosate blocks the biosynthesis of aromatic amino acids. The introduction of a gene that codes for a modified form of the enzyme EPSP (5-enolpyruvyl-shikimate-3-phosphate synthase) results in resistance of the herbicide glyphosate.

Overproduction of Proteins

The introduction of several copies of a gene or the use of a strong promoter can result in the overproduction of the protein. This can be used in some nutritional applications or for disease resistance when certain proteins are needed to fight disease.

Silencing of Endogenous Genes

Partial or total suppression of gene expression can be obtained by RNA antisense technology. This technology consists of the introduction of a gene in the reverse orientation in relation to the original sequence. When transcribed, this produces a polynucleotide complementary to the original gene. The mRNA of the gene of interest is complementary to the introduced gene, resulting in the formation of a double helix RNA that blocks the translation process. Theoretically, the antisense mRNA can be used to inhibit the expression of any gene.

New Traits

Genes of other species can be introduced in the target species, making traits available in one species available to any other. Several possibilities exist, including these:

- Metabolism: Transfer of genes from nitrogen-fixing species to grasses.
- Biopesticides: The Bt gene was transferred from the bacterium *Bacillus thuringiensis* to corn, cotton, and other crops.

Disease Resistance

An example is barley resistance to Barley Yellow Dwarf Virus (BYDV), which resulted from the introduction of the gene that codes for the envelope protein of BYDV into barley.

Male Sterility

The introduction of male sterility genes can increase the rate of cross-fertilization in a self-pollinated species.

Bioremediation

Introducing genes that code for the capability of absorption of heavy metals or the capability of metabolizing pollutant residues could have important applications in biodegradation. This subject is also discussed in Chapter 11, "Bioremediation."

Pharmaceuticals

The introduction of genes that code for production of substances with therapeutic properties can be used for the production of medicines.

Alteration in the Individual's Architecture

Altering flowering time, plant architecture, or coloration might have important applications in ornamental plants.

GENE EXPRESSION . . . .

All cells possess the typical number of chromosomes of their species. Therefore, root, epidermis, or pod cells of a soybean plant possess all 40 chromosomes typical of this species. However, not all of the genes are expressed in every cell. For instance, genes that code for chlorophyll production are expressed in the leaves and any other green part of the plant. However they are silenced in the roots, which is the reason these cells do not contain chlorophyll. Gene regulation is a complex process that is affected by a series of factors. A common occurrence in genetic engineering is a lack of expression after a gene has been transformed into an organism. Therefore, an understanding of mechanisms involved with gene expression is critical in genetic transformation.

In bacteria, some genes are activated while others are silenced, depending on the conditions in which these microorganisms are grown. For example, the bacteria *Escherichia coli* can use two different carbohydrates, lactose and glucose, as energy sources. The bacteria needs to synthesize specific enzymes that catalyze the breakdown of the carbohydrates into energy. The enzymes, like all other proteins, are coded by genes. When *E. coli* is cultivated in a medium with both glucose and lactose (preferably glucose), it metabolizes. The genes coding for the production of the enzymes that metabolize glucose are thus expressed preferentially. The metabolism of lactose requires an additional enzyme that is only synthesized, or activated, after the medium runs out of glucose and lactose is the only energy source available. This phenomenon is called *gene regulation*.

Gene expression in more complex organisms is still not completely understood. The complexity of gene regulation is a puzzle in the zygote, a cell formed by the union of sperm and egg cells, in which the genes coding for differing functions have to be activated in a precise and orderly manner. The same genetic information present in the zygote is also present in any other cell in the body, from muscles to skin. Obviously, different genes are activated or expressed in each organ in a different way.

Gene expression is not just a function of where the cell is, but also the result of environmental stimuli. Cells of a floral bud of soybeans differentiate into flowers when the plant is grown during long nights. If the soybean plant is grown during short nights, it continues vegetative growth and does not bloom. Another example of gene regulation occurs with animals, including humans. Testicle and ovary cells do not start the production of sexual hormones until the individual reaches puberty.

Another example of the complexity and importance of gene regulation can be observed in the metamorphosis and development of butterflies and moths. These insects take three forms during their lives: caterpillar, pupa, and adult butterfly or moth (Figure 3-7). The insect possesses the same genes and DNA during these three different developmental phases. Although the caterpillar, pupa, and adult have the same genes, it is interesting to observe that different genes

Figure 3-7
Expression of different genes during the development
of the corn earworm moth.
Source: Courtesy USDA-ARS. Bottom photo by Juan Lopez.

are expressed in the three developmental phases. In the caterpillar
phase, the genes for production of several legs and a stronger mouth
capable of chewing leaves are expressed, but not the genes for production of wings. However, the genes for the formation of a delicate
mouth apparatus, appropriate for nectar feeding, and genes for the
formation of wings are active in the insect's adult phase. The gene expression pattern changes during insect development to allow for the
correct progression of its life cycle.

The mechanisms regulating gene expression involve regulatory genes. As opposed to the genes discussed up to this point, these DNA sequences do not code for any protein. Their function is to promote the activation or the silencing of genes.

An important part of gene regulation is the promoters. A promoter is a DNA sequence preceding the gene, which contains regulatory sequences to control the rate of RNA transcription. Promoters control when and in which cells a certain gene is expressed. Through the manipulation of promoters it is possible to induce superexpression, underexpression, or even gene silencing.

Some promoters are constitutive—that is, they induce gene expression continually—whereas others are inducible. Among these, there are some that are chemically inducible, and others are activated by heat, light, or hormones. Some promoters are active in certain tissues and organs, but not in others. In this case, they are considered tissue-specific promoters, as in the case of chlorophyll production. The promoters of the chlorophyll genes are not active in roots, but they are active in the leaves and in all green parts of plants.

Some of the promoters frequently used in genetic engineering of plants include the following:

- Constitutive
  - UBI from corn
  - 35SCaMV from a cauliflower virus
- Tissue-specific
  - Phaseolina promoter, a seed-specific promoter from field beans
  - Vicillin promoter, a seed-specific promoter from peas
  - Glutamine promoter, an endosperm-specific promoter from wheat
- Inducible
  - Rubisco 5S promoter, inducible by light

Aside from promoters, other genetic factors are important in proper gene expression. Although the genetic code is universal, it is also considered degenerate, as more than a single codon codes for a certain amino acid (see Chapter 1, "History: From Biology to Biotechnology"). Different organisms have acquired the preferential use of specific codons for certain amino acids during evolution; this can also have an impact in gene expression. That was the case of the Bt gene from *Bacillus thuringiensis* introduced in corn. Initially, the expression of that bacterial gene in corn was low; however, when a transgene was reengineered to favor the preferential use of certain codons by corn, gene expression occurred at normal levels.

Several other factors can affect the expression of transgenes, such as the presence of a peptide signal, the site of its integration in the genome, the number of copies integrated, and transgene rearrangements during the integration process. Integration of transgenes in the host genome, in general, happens at random; that is, it can occur in any chromosome of the cell and it can land in any part of the chromosome. However, most of the transgenic varieties have the transgene inserted close to the ends of the chromosome. Multiple copies of the transgene are typically introgressed together.

TRANSGENIC LOCUS . . . .

Gene constructs used in genetic transformation posses a promoter, coding region, and termination sequence (Figure 3-8). In Figure 3-8, the vicillin promoter, specific for expression in seeds, drives the expression of the gene UDP 6-glucose dehydrogenase in the antisense orientation. The construct also possesses the NOS (noplaine synthase) termination sequence, which marks the site for the end of transcription. Besides the transgene of interest, in general, reporter genes are introduced simultaneously to facilitate the identification and selection of transformed individuals.

For the selection of the transformed cells, the gene construct contains a gene sequence that codes for antibiotic or herbicide resistance. Frequently, neomycin or hygromycin antibiotic resistance genes or

Figure 3-8
Example of the genetic construct used for silencing the gene UDP 6-glucose dehydrogenase in soybeans.

the phosphinotricim acetyl transferase herbicide tolerance gene is included under a strong constitutive promoter, such as 35SCaMV. The transformed cells would be the only ones possessing the capability to grow in a medium with a selective agent (antibiotic or herbicide), thereby facilitating their selection.

Frequently, a gene reporter is also included in the genetic construction. The function of this reporter is to allow the visual identification of transformed cells. Three genes have been used as reporters in plant transformation: Glucaronidase (GUS), Luciferase (LUC) and Green Fluorescent Protein (GFP). GUS allows the identification of transformed individuals by the expression of a blue color, because they become blue in the presence of the chemicals X-Gal and IPTG (isopropyl beta D-thiogalactoside). Luciferase, a protein in fireflies, turns the transformed individuals phosphorescent, and GFP, isolated from a species of jellyfish, codes for a fluorescent-green color in transformed individuals.

Genetic transformation of individuals is a difficult task. The science behind the methods is understandable on a basic level, but the results from the procedures do not always work out as planned. Specific gene sequences are needed to induce the expression of a transgene, and genes are needed to identify the transformed cells. Still, the use of transformation is being improved to more accurately express desired traits in different organisms. The comprehension of the intricacies of transformation is a key to understanding the broad applications of biotechnology.

4 Biotechnological Products

In this chapter...

Some of the common questions of agricultural science during the 1980s and the beginning of the 1990s were these:

1. Will biotechnology be the solution to world hunger?
2. Will it substitute for conventional genetic improvement?

For those at the forefront of this new science, these questions resounded with encouragement and challenge. For others who did not feel part of the new science, the questions seemed to threaten conventional methods. These decades have passed and the predictions made then can today be compared in the fields of the farmer.

Although many forums for debate feature people in favor of this technology, many are against its use. The debate about the risks and benefits of this biological technology continues to ignite heated arguments. Various philosophers believe this polarization of society has many beneficial aspects, for it warns scientists of the need to be vigilant in constantly reassessing possible risks.

Although many have made predictions that biotechnology would resolve agricultural problems and end world hunger, some skeptics believed that this science would not be able to generate commercial products that would be competitive in a worldwide market. Perhaps a balanced assessment of the first decades of agricultural biotechnology would show that the predictions from those most optimistic and the most skeptical were both wrong.

Today, society is a witness to the fact that biotechnology develops products and services that have a great impact in the lives of farmers and consumers in many countries. Transgenic varieties of soybeans, corn, cotton, canola, papaya, rice, tomato, and other species have captured the preferences of farmers since the first transgenic variety, the Flavr Savr tomato, was released in May 1994. The agricultural census of 2000 records that worldwide, more than 44 million hectares were planted with transgenic crop varieties, showing a continued increase in the area occupied by these varieties. Soybeans, for example, currently have transgenic varieties planted on 36 percent of the soybean acreage worldwide.

In animal production, the first lines of transgenic species came more slowly. The first animal commercially available on a large scale was the salmon. Transgenic salmon reached American markets in

2001, following rigid evaluations of consumer and environmental safety. Other transgenic animals for food production are in their final steps of development.

Even though many transgenic products are on the market, there is no doubt that biotechnology has neither resolved all of the world's agricultural problems nor abolished world hunger. An understanding of the complexities of agriculture and the factors contributing to world hunger help one understand that biotechnology is only one of the technologies that is able to contribute to the well-being of society. This science constitutes only one part of an integrated solution for the many challenges facing farmers in the production of high-quality food in adequate quantities to meet global demand.

ECONOMIC ASPECTS . . . .

Agriculture constitutes a major sector of the global economy. The production and commercialization of food, feed, fiber, and other agricultural products influences the movement of the majority of the world's resources either directly or indirectly. Agribusiness has an impact on the life of all inhabitants of the planet, for no human being is able to feed or clothe himself or herself without agriculture.

The applications of biotechnology in agriculture have a potential market estimated at $67 billion per year. Because of this, it is not surprising that many multinational companies have economic interests in this new science. The competition for this market has motivated these companies to dedicate a great part of their resources to research on and development of new products. Various groups are working on the development of transgenic animals and plants with greater productivity, resistance to diseases and pests, improved nutritional quality, and better tolerance to environmental stresses.

TRANSGENIC ANIMALS . . . .

One of the first genetically modified products available for animal production was the bovine somatropin hormone (BST), also called bovine growth hormone (BGH). This hormone is naturally produced in the pituitary gland of the brain. It promotes the growth of young animals and regulates milk production in dairy animals. The genetically modified hormone has been used to increase milk production up to 20 percent in dairy cows. This hormone is produced by transgenic bacteria, in which the bovine gene that codes for BST was inserted into their genome. Therefore, the hormone given to cattle is essentially the same as the one produced in the pituitary gland of the animals.

BST was approved by official organizations of the U.S government and released for use in 1994. Since then, farmers have been able to stabilize milk production in their herds, avoiding inconvenient fluctuations in production levels. Naturally, dairy cows reach peak milk levels around 50 days after calving, after which levels decline during the following 10 months. The use of BST allows the farmer to provide a more constant milk volume.

This technology has provoked debate and worry among many. The public has raised various questions about the economic advantages and risks to both human and animal health. Some consider inducing increased milk production both cruel and harmful to the health of the animal. Although opponents of BST argue that the hormone results in animals with brittle bones, mastitis (an infection of the mammary glands), and decreased resistance to diseases, these problems have not yet been scientifically confirmed. In fact, no known side effects from this hormone to animals or humans have been detected to justify its removal from the market. Another argument against BST is that any risk to human health cannot be measured by tests on short-lived laboratory animals. Only long-term consumption of milk from animals treated with the hormone is able to generate a more conclusive evaluation. Some nongovernmental organizations (NGOs) oppose the use of BST on the basis of ethical principles. These organizations consider it unethical to use a hormone to increase milk production to the maximum limits of production, as if the animals were simply machines.

The U.S. Food and Drug Administration (FDA), an agency that evaluates and authorizes the release of drugs in the country, analyzed questions about the risks of BST to human health. The FDA's conclusion following a detailed study was that the milk and meat from animals treated with BST are safe for human consumption.

Today, BST is legally used for milk production in many countries around the world.

The lack of evidence against BST and its increased use in different countries has caused NGOs to change their tactics. Instead of proposing a ban on the use of the hormone, their strategy has shifted to support the labeling of milk produced with BST.

Among the promises of biotechnology in this area are genetically modified cows that produce, in their own glands, a greater quantity of BST, thereby eliminating the need to give the hormone to animals during lactation.

Especially in animal production, biotechnology can contribute to improvements in the quality of meat, milk, eggs, and wool, as well as disease resistance in animals, which would serve to reduce the use of antibiotics in production processes. As mentioned earlier, salmon was the first transgenic animal approved for human consumption. This animal possesses an additional gene that codes for a growth hormone that allows the fish to grow more rapidly because of an improved food conversion rate, which is 15 percent greater than it is in nontransgenic salmon (Figure 4-1).

The transgenic salmon were obtained through genetic transformation of the sperm from male fish, which were then used to fertilize the female eggs, thereby producing a transgenic zygote (Figure 4-2). The primary methods of reproductive cell (sperm and egg) transformation in fish are electroporation, microinjection, and biolistics (see Chapter 3, "Transformation").

It is predicted that within a few years, there will exist transgenic lines of poultry, producing eggs that will have a lower content of cholesterol; cows producing milk with lower lactose levels, which is extremely important for some ethnic groups; lines of sheep with longer and stronger wool fibers; and more. Although animal breeding has been able to make substantial progress for some of these

Figure 4-1
Genetically modified salmon, the first transgenic animal for food production.

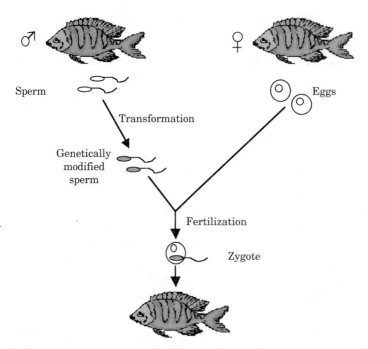

Figure 4-2
Production of transgenic fish.

traits, some of the promises might only be realized with the help of biotechnology. One example is pork production, where fat content must be reduced. Biotechnology might be the best way to significantly reduce fat. This is especially important when the desired trait is not present in other animals of the same species or in species that are sexually compatible. Biotechnology allows the transfer of genes, and therefore traits, among different species.

Actually, there are transgenic lines of sheep, goats, and swine with different transgenic traits in the final phases of evaluation. These animals might be useful, not only for food production, but also as "bioreactors" in the production of hemoglobin and antibodies to combat snakebites, as well as other applications in pharmaceuticals, nutrition, and the production of tissue and organs for human transplants. For example, during the 1980s, some hemophiliacs were infected with the AIDS virus because blood banks failed to test donors for HIV. Although the blood banks continued the use of human blood, two problems became evident: the problem of reduced donations not meeting the growing demand, and the risk of transmitting contagious diseases for which testing had not been developed. One project that seeks to address these problems is the development of transgenic swine that produce human plasma and hemoglobin. Presently, there is no evidence of HIV in swine, making them potential blood donors. Even so, tests are being done to learn about the possible transmission of other diseases and viruses in swine to humans.

Transgenic Animals as a Source of Organs and Tissue

Xenotransplant, or tissue or organ transplant among different species, genera, or families, has been indicated as one possible solution to the shortage of donor organs. A common example of xenotransplant in use today is the use of swine heart valves in humans (see Figure 4-3). An additional problem with transplantation is rejection. When cells or tissue from an individual are transplanted to another individual, the recipient recognizes the transplanted tissue as being foreign (an antigen). This begins the production of antibodies that attack the

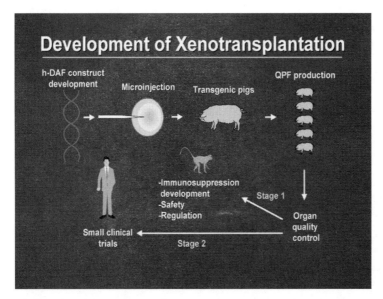

Figure 4-3
Development of transgenic animals for xenotransplant.

antigen, causing rejection of the organ. This situation could be remedied with the use of transgenic animals.

With the aging of the population, the number of patients needing organ transplants has been increasing significantly, but the number of donors has been growing only moderately. In spite of the preventive campaigns that emphasize the importance of controlling blood pressure and cholesterol levels, maintaining a balanced diet, and eliminating sedentary habits and cigarette use, cardiovascular diseases continue to kill four times more people than AIDS, and three times more people than breast cancer. A more effective therapy for many heart disease patients is the use of organ or tissue transplants. Approximately 45,000 Americans under age 65 could benefit from a heart transplant, but only 2,000 human hearts are annually available for transplant in the United States.

Xenotransplantation might eventually be used not only for heart patients, but also for the treatment of kidney diseases, diabetes, Parkinson's disease, Alzheimer's disease, and even in the treatment of third-degree burn victims.

Recently, scientists have created genetically altered pigs with a gene involved in the rejection of transplanted organs. Although this is still an experimental procedure, xenotransplants are extremely promising from the perspective of the great demand and limited availability of organs, and the resultant patient waiting lists.

The first transgenic pigs (Figure 4-4) developed for xenotransplants by PPL Therapeutics in December 2001 have the gene alpha

Figure 4-4
Transgenic pigs developed for xenotransplant.

1,3 galactosyl transferase silenced. This gene codes for an important enzyme in sugar metabolism in the swine's cell membrane.

Swine have received the most attention from scientists in the field of xenotransplantation because they are so biologically similar to humans and they are not carriers of human diseases. Primates, although genetically more similar to man, involve additional risks in the transplant cases because they are known carriers of human diseases.

Some scientists, however, are aware of the indirect risks of xenotransplantation. This technique might transfer viruses or other infectious microorganisms from animals to humans. One example is the porcine endogenous retrovirus (PERV), a virus that has evolved with swine but does not cause disease. It is not known, however, what might happen if these nonpathogenic viruses were introduced into humans through transplanted organs. Additionally, ethical issues have also been raised concerning the implications of people using animal parts in organ or tissue transplants.

TRANSGENICS IN AGRICULTURE . . . .

The world population reached 6 billion in 2000; this signifies a doubling of the population in the last 40 years. There is no global initiative to reduce population growth, and the expectation is that the world population will reach 9 billion inhabitants around the year 2040, generating a 250 percent increase in demand for food.

An increase in food production might come from the following three areas:

1. Adding new land to production. This option is extremely unlikely or even impossible in many developed countries where all available land is being used for agriculture. In the United States, Brazil, and a few other countries, the expansion of agricultural frontiers is still possible where large tracts of native land still exist, but are not currently being exploited. However, this poses the threat of significant harm to the environment.

2. Reducing losses, whether they are caused by pests or diseases, or from cultivation or harvesting problems. The losses from transportation and storage also contribute to waste of food.

3. Increases in crop productivity.

Biotechnology can contribute in each of these three areas to increase food production. In what has been said about expanding agricultural frontiers, biotechnology can contribute crop varieties with more tolerance to drought or poor soil conditions. Pests, diseases, and weeds not only reduce agricultural productivity; they also contribute to increased production costs, as they require farmers to apply agro-chemicals. Another inconvenience associated with agro-chemicals is the environmental pollution and chemical residues on produce. There exist about 40,000 species of microorganisms that cause diseases in plants and approximately 30,000 species of weeds that compete with crops for nutrients, water, space, and light, causing additional reductions in productivity. The first transgenic plant variety produced on a large scale was developed to assist in the control of weeds, as is the case in soybeans resistant to the herbicide glyphosate.

Varieties Resistant to Herbicides

The difficulty in controlling weeds through the application of herbicides is that there is no single chemical product designed for broad-spectrum weed control that does not cause injury to the crop. Normally, when an herbicide is effective in controlling weedy grasses, it is not effective on broadleaf weed species. The development of crop varieties tolerant to herbicides has been a major focus for biotechnology companies. Today there are varieties of various crop species with resistance to nonselective herbicides, or herbicides that kill all types of plants.

The idea of developing crop varieties with tolerance to herbicides came from an observation that some weeds acquired tolerance to some chemicals after repeated sprayings. Scientists found that some herbicide-resistant weeds or even bacteria have altered enzymes or other mechanisms to inactivate the herbicide. This information led to research on using biotechnology to transfer the modified enzymes to crops, which confers resistance to the herbicide. Monsanto developed the first transgenic varieties tolerant to herbicides. These varieties are known as Roundup Ready, as they are resistant to the herbicide glyphosate, which has the trade name Roundup. Glyphosate blocks the synthesis of the aromatic (ringed) amino acids by binding and inactivating the enzyme EPSPS, which is essential in the synthesis of these amino acids in plants. Roundup Ready crop varieties include a gene that codes for an altered form of the EPSPS enzyme that has less affinity to the herbicide, but is still able to catalyze reactions and synthesize aromatic amino acids. The gene for resistance to glyphosate has already been introduced into many crop species of economic importance beyond soybeans.

Farmers have shown a great interest in planting Roundup Ready varieties because they allow for more efficient and economic weed control with application of only one herbicide. With the use of one herbicide to control the majority of weeds, farmers can reduce the number of herbicide applications needed for effective weed control.

Considering that varieties tolerant to herbicides would reduce the amount of chemicals applied to the land, they should also be preferred by consumers. However, a lack of communication about the advantages of these varieties not only for the farmer, but also for the environment and consumers, has divided public opinion, creating questions about the safety of these crops. Even though these varieties, as with any other genetically modified organisms, pass through stringent biosafety evaluations that analyze the risks to human and animal health and to the environment (tests of environmental impact), many people continue to be skeptical about the benefits of transgenic crops.

Crop Varieties Resistant to Insects

One of the most effective biocontrol agents is the bacterium *Bacillus thuringiensis* (Bt). This bacteria has been used in the control of caterpillars in fields since 1980, when it was discovered that the bacteria produces a protein that is toxic to the lepidoptora species of insects, which includes caterpillars. When caterpillars eat plant leaves on which the bacterial spores have been deposited, the bacteria grows in the digestive tract and a protein-like crystal is produced (Bt protein). The protein basically attaches to cellular membranes of the caterpillar's digestive system, altering osmotic potentials and eventually killing the insect. The digestive system in mammals, including humans, produces an acid that rapidly degrades the Bt protein if it is ingested, making it harmless to human health. The Bt insecticide has been considered one of the safest chemicals to man and the environment. Because of its effectiveness, scientists wanted to create transgenic plants that were able to express the Bt protein.

This gene has been transferred to varieties of corn, cotton, tobacco, potato, and other species to extend the control of caterpillars. Bt crop varieties synthesize the Bt protein in their own tissue. After eating plant matter (leaves, stems, etc.) from Bt varieties, caterpillars die within two days. This has dramatically reduced the amount of insecticide needed to control the insect in crops.

However, some people still prefer to use products from conventional corn, produced under an intense program of chemical insecticide applications, instead of those obtained from Bt corn.

One of the worries with the use of Bt corn is the development of resistance within the insect population and a subsequent loss of efficacy in pest control with the Bt protein. Because a plant is continually producing the Bt protein in its tissues, there is an increased chance that insect populations will develop resistance. To address these potential problems, farmers who adopt this technology are required to implement a resistance management plan that establishes a refuge area where the insects can reproduce and feed on conventional plants that do not produce the Bt protein. Additionally, farmers are encouraged or, at times, requested to have a rotation of conventional cultivars with the genetically modified cultivars. These practices reduce the selection pressure on the insects, thereby reducing the likelihood of resistance developing in the population.

Plant Varieties for the Production of Bioplastics

Biotechnology has the potential to improve not only food production, but also nonfood products such as plastic (see Figure 4-5). Plastics are long polymers based on organic compounds. Although the major component of plastic has petroleum as its primary material, it can also be synthesized in plants. Currently, an interesting project is

Figure 4-5
The replacement of plastic derived from petroleum with
bioplastic would reduce environmental pollution.

underway at the University of Minnesota aiming to develop transgenic crop varieties for bioplastic production. The bacteria *Alcaligenes
eutropus* handles the production of polyhydroxybutyrate (PHB), a
biodegradable and renewable biopolymer (independent of
petroleum). The gene from *A. eutropus* that codes for an enzyme responsible for the biosynthesis of PHB is being transferred to corn, a
species highly efficient in biomass production, which will contribute
to reductions in the production costs of this biodegradable plastic.
Obtaining transgenic species with an elevated expression of this gene
would certainly make biodegradable plastic more competitive in the
marketplace.

Corn is a highly productive and efficient plant species. Consider
the case of cornstarch, which is produced at a similar cost to
petroleum for each unit of energy generated. However, contrary to
petroleum, plant-based energy sources are renewable and more envi-
ronmentally friendly. Through engineering the genes that code for
cornstarch, it is possible to produce starch with different properties.
For example, scientists are developing an edible bioplastic that
would allow foods to be cooked in their own packaging, thereby reducing the volume of domestic waste.

Plants and Animals as Bioreactors

Rural enterprises will undergo substantial changes in the coming decades. Grain collected in the field, milk from cows, and eggs produced from chickens will probably have other fates than merely serving to feed the population. Genetic engineering will produce varieties of plant species and lines of animals that will function as bioreactors, or living factories for the production of pharmaceuticals, chemical products, plastics, fuel, and other products. Because of this, farms will stop being only a source for food.

Plants already produce a variety of chemicals used by industry for production of medicines (*Pilocarpus pinnafolius*), dyes (*Bixa orellana*), paint (soybeans), and industrial oils (canola). The introduction of new genes could alter the quality and quantity of existing products. For example, altering the composition of fatty acids in legumes such as soybeans, canola, and peanuts, it is possible to develop different types of oils that can be used for healthier diets for people with cardiovascular diseases or for the production of hydraulic fluids for automotive and industrial uses.

A curious example that is still found in studies is the production of anticoagulants in the canola plant. In this project, scientists isolated genes that code for the production of anticoagulation factors from a leech, which were then transferred to canola. This anticoagulation factor is synthesized by the plant and stored in the seeds, allowing relatively easy extraction and purification of the factor. These anticoagulants have important applications in the treatment of circulatory diseases.

Cattle, swine, sheep, and poultry have been genetically modified to produce different proteins with applications to human health. This form of protein production introduces a series of advantages over traditional methods, including lower costs and reduced risks of contamination. The use of tissue-specific DNA promoters, meaning promoters that activate transgenes only in the desired organs or tissues, reduces the risks of side effects for the transgenic animals. For example, the use of a promoter that only expresses a transgene in the mammary glands induces the animal to deposit the protein in the milk, thereby simplifying collection and purification.

Another example is the production of antibodies. An egg yolk normally contains antibodies that are deposited by the hen to protect the

Figure 4-6
ANDi, the first transgenic monkey obtained in the Primate Research Center in Oregon.
Source: Courtesy of Oregon Health and Science University.

embryo from infections in the period preceding the development of its immune system. The types of antibodies deposited can be modified through immunization with specific antigens. With transgenic poultry there is no need to immunize the birds for the production of eggs with antibodies to combat diseases of other animals, including humans.

Finally, transgenic animals developed for pharmaceutical studies also have great importance. For example, a line of transgenic mice in which researchers inserted an oncogene (a gene related to cancer formation) was the first animal to be patented. Researchers have used these mice to study different cancer treatment therapies. However, the first transgenic monkey was obtained with the introduction of the GFP gene, by way of a retrovirus (Figure 4-6). This animal has been named ANDi, an acronym for inserted DNA, written backwards. ANDi has been important in more advanced biological and genetic studies.

FINAL CONSIDERATIONS . . . .

Transgenic products are already available in many countries. Defenders of biotechnology have emphasized the benefits of these products to society. The potential realized to this point is only a small glimpse in relation to what will be realized in the near future. However, those opposed to biotechnology have presented a list of concerns that have strong appeal to society. Some of the arguments against transgenics include the following:

- It only serves the interests of large corporations.
- It does not favor sustainable agriculture.
- Biotechnology only benefits large farmers.
- It creates dependence on other products and services from large multinational companies.

Amidst these critics, the scientific world hopefully awaited a product of genetic engineering that would not be the target of any of these arguments. Finally in 2000, the Swiss Institute of Crop Science in Zurich, a governmental agency, released a transgenic rice variety dubbed Golden Rice. This rice (Figure 4-7) rapidly gained the sympathy of society and attracted international media attention. This transgenic variety was the result of work by a group of Swiss and German researchers, through funding from the Rockefeller Foundation, the European Community, and the Swiss Institute of Technology. Golden Rice consists of rice lines that have elevated levels of β-carotene, a precursor to vitamin A. Within this rice plant, the biochemical pathway for β-carotene was engineered by inserting genes from the daffodil (*Narcissus pseudonarcissus*) and a fungus (*Erwinia uredovora*). This rice variety was developed to help combat blindness resulting from a deficiency of vitamin A, a critical problem in less developed countries, such as some in Africa and Southeast Asia.

With general approval from the public, even the less optimistic organizations, notably those critical of genetically modified organisms (GMOs), were expected to approve the arrival of Golden Rice. To the

Figure 4-7
Grains of Golden Rice, rich in β-carotene.
Source: Courtesy of Swapan Datta, International Rice Research Institute (IRRI).

surprise of many, some NGOs are trying to prevent Golden Rice from being made available to the small farmers of countries where vitamin A deficiency blindness is a major problem. Golden Rice was not developed by a multinational company, does not require the use of other agro-chemicals, was not developed for large farmers, and is not incompatible with sustainable agriculture. How is it possible that this product could arouse so much opposition? Unless these organizations are failing to exercise their critical senses, how can one understand such rejection amidst the reality from UNICEF data showing that 1 million to 2 million deaths could be avoided annually if a vitamin A supplement were available? Additionally, 1,369 children lose their vision daily because of vitamin A deficiency. Many still question why there has been so much opposition to the release of a product with such humanitarian potential.

It is utopian, however, to think it possible to have reconciliation between groups with such contrasting ideologies. It would be irresponsible to argue that all the biotechnological products represent only benefits without risks. It is important that society, with its many forms of expression, including NGOs, should be vigilant and question technological advancements. Even with ideological differences, there exist common objectives among all people, an example being the production of abundant food with high nutritional quality at accessible prices with minimum damage to the environment. Although it might be true that businesses envision as a priority products that bring them economic return, it is also true that they are aware that society will not tolerate products that are harmful to the environment or human health. Although environmentalists consider the multinational companies their enemies, they are blind to the fact that these companies are also allies in the search for alternatives and solutions to the problems facing humanity.

5　Biosafety

$\mathbf{B}$iosafety is a science that began only in the last century. It is concerned with the control and minimization of risks resulting from biological technologies. This area of science studies the impacts of biotechnology on human and animal health as well as the environment. The diverse fields of chemistry, law, public policy, health, and even agriculture are integrated within the realm of biosafety. On a global level, biosafety is regulated in different countries by specific laws.

In the last three decades, greater environmental concerns have been voiced worldwide, mainly due to increases in air and water pollution from industry and transportation. Estimates from the World Bank suggest that about 20 percent of the world population does not have access to clean water. There has never been as much demand for ecological preservation as there is now (see Figure 5-1).

It was in this environment that genetic engineering appeared in the early 1970s in California. The initial experiments in 1973 elicited a strong reaction from the scientific world, and culminated with the Conference of Asilomar in 1974. At this conference, the scientific community effectively proposed a moratorium on the use of genetic

Figure 5-1
Some NGOs have been organizing popular demonstrations against genetically modified organisms in several countries.
Source: Courtesy of John Maihos, boston.about.com.

engineering until proper mechanisms for safety were established. Scientists wanted to guarantee that biotechnology could be used with minimal risks to humans and the environment. In a relatively short period, scientists developed biosafety rules to govern the use of biological technologies. In almost 30 years of biotechnology studies, there have been no documented direct cases of harm to human and animal health or to the environment from genetic engineering. Biosafety is concerned with the control and minimization of risks from the practice of different technologies. Biosafety aims to study, monitor, and control the potential impacts of biotechnology.

Biosafety legislation should include all biological technologies and not only genetic engineering or recombinant DNA technology. It should also establish requirements for research, handling, and commercializing of GMOs. The basic foundation of biosafety is to protect human and animal health and the environment while assuring the progress of biotechnology.

Several products developed from recombinant DNA technology are being marketed worldwide. In North America, the following transgenic products are now available: papaya, corn, soybean, cotton, canola, human insulin, BST, salmon, and more. Biotechnology regulation around the world has been carried out with sound science, transparency in decision making, consistency and fairness, collaboration with regulatory partners, and the building of public trust. The need for federal regulation or oversight of GMOs became apparent in the mid-1980s, as companies asked for clarification of existing regulations as they related to genetically engineered microorganisms and plant pesticides.

U.S. REGULATORY AGENCIES . . . .

In the United States, the agencies that examine plants and plant products are the Environmental Protection Agency (EPA), the FDA of the U.S. Department of Health and Human Services, and the Animal and Plant Health Inspection Service (APHIS) of the U.S. Department of Agriculture (USDA).

The USDA, through APHIS, regulates the development and field testing of genetically engineered plants, microorganisms, and certain other organisms under the authority of the Federal Plant Pest Act and the Plant Quarantine Act. APHIS regulations provide procedures for obtaining a permit or for providing notification of the intent to field test prior to importation, interstate movement, or release into the United States. Permission is granted based on the genes involved and the plant pests controlled by the transgenes.

APHIS has been reviewing applications for permits and notifications by industry, academia, and nonprofit organizations for field tests of transgenic crop plants since 1986, when it first proposed regulation of these products. After several years of field tests, an applicant can petition the agency to be released from requirements under the APHIS regulatory process. If the applicant can provide evidence that there is no plant pest potential (including the lack of change in disease and pest resistance, as well as the absence of the potential for new genetic material to create a new pathogen or pest), along with answers to a variety of other environmental questions, APHIS will grant the petition. At that time, the applicant is free to commercialize the plant line or use it in other breeding programs without going to APHIS for permission, subject to any necessary approvals from the EPA or FDA. To date, 51 petitions have been granted and more than 5,000 permits and notifications have been issued for field testing at more than 22,000 sites. Although no petitions have been denied, 13 have been withdrawn due to insufficient information or other application inadequacies.

APHIS maintains comprehensive field testing and petition databases that are used not only by domestic customers and stakeholders, but increasingly by foreign governments to verify that the U.S. government has reviewed the risks associated with products being considered for field testing or importation. These databases, as well as access to federal home pages on biotechnology regulation, are available at *www.aphis.usda.gov*.

The EPA's responsibility is to ensure the safety of pesticides, both chemical and biological, under the authority of the Federal Insecticide, Fungicide, and Rodenticide Act (FIFRA) through regulation of the distribution, sale, use, and testing of plants and microbes producing pesticidal substances. Under the Federal Food, Drug, and Cosmetic Act (FFDCA), the EPA sets tolerance limits for substances used as pesticides on and in food and feed, or establishes an exemption from the requirement of a tolerance if such a tolerance is not necessary to protect the public health (determined after evaluation by the agency).

The EPA issues experimental use permits for field testing of "pesticidal" plants and registrations for commercialization of these plants. The *Bacillus thuringiensis* (Bt) toxin, which occurs naturally in soil bacterium, is considered a biological pesticide. For plants containing Bt toxin, the manufacturer must prepare a resistance management plan as a condition for registration with the EPA. The plan has to describe how the manufacturer registering the plant product will assure that resistance does not build up in affected insect populations and reduce the effectiveness of Bt applied topically or used through the plant's genetics. The EPA also evaluates the new use of herbicides on herbicide-tolerant transgenic plants.

The FDA assesses food (including animal feed) safety and nutritional aspects of new plant varieties as part of a consultation procedure published in the 1992 Statement of Policy: Foods Derived From New Plant Varieties. Consistent with its 1992 policy, the FDA expects developers of new plant varieties to consult with the agency on safety and regulatory questions under the authority of FFDCA. FDA policy is based on existing food law and requires that genetically engineered foods meet the same rigorous safety standards as all other foods. The FDA biotechnology policy treats substances intentionally added to food through genetic engineering as food additives if they are significantly different in structure, function, or amount from substances naturally found in food. Many of the food crops currently being developed using biotechnology do not contain substances that are significantly different from those already in the diet and thus do not require premarket approval.

The EPA's jurisdiction under FIFRA is limited to pesticidal substances. For example, a plant that has been genetically modified to resist disease comes under FIFRA authority, whereas a plant that has been modified to resist drought does not. The former comes under EPA authority because the substance produced by the plant acts as a pesticide by affecting a pest. In the latter instance, a substance produced by the plant might result in, for example, deeper roots to enable the plant to access more water reserves. This transgenic plant would be subject to USDA regulation, and any food or feed produced would be subject to FDA authorities.

For plant pesticide varieties, the EPA has four categories: product characterization, toxicology, effects on nontarget organisms, and exposure and environmental fate. Product characterization includes reviewing the source of the gene and how the gene is expressed in a living organism, the nature of the pesticidal substance produced, modifications to the introduced trait as compared to that trait in nature, and the biology of the recipient plant. For toxicology, an acute oral toxicity test of the pesticidal substances on mice is required. At times, it has not been possible to make enough of the substance in the plant itself, so the EPA has allowed the exact same protein to be produced by bacteria and used for the testing. It should be noted that to date, all of the plant pesticides reviewed by the EPA are proteins plus the genes required to make these proteins within the plant. For these proteins, the EPA also requires a digestibility test to determine how long it takes for the protein to break down in gastric and intestinal fluids. The EPA also considers the potential of allergenicity. Determination of whether an introduced protein is likely to be an allergen is one of the major challenges for the federal agencies. The EPA and FDA work on this issue together.

For ecological effects, the EPA examines the exposure and toxicity of the plant pesticide to nontarget organisms, such as wildlife and beneficial insects. These tests are unique to the crop and pests involved. For example, during the review of the plant pesticide Bt-potato, a test of potential effects of the introduced protein to ladybird beetles was conducted and showed that there were no adverse effects to these predators or the pesky Colorado potato beetles. For Bt-corn, tests were conducted on the potential effects on fish because field corn can be manufactured into commercial fish food. No effects were observed in the tests. EPA also has evaluated the degradation rates of the proteins in soil and plant residues.

BIOSAFETY IN OTHER COUNTRIES . . . .

Biosafety regulation is carried out in different countries by local agencies. In Canada, this regulation is administered by Health Canada, the Canadian Food Inspection Agency, and Environment Canada (EC). Together, these three agencies monitor development of plants with novel traits, novel foods, and plants or products with new characteristics not previously used in agriculture and food production. In Japan, all products derived from plant biotechnology are subject to rigorous testing procedures overseen by the Ministry of Agriculture, Forestry and Fisheries and the Ministry of Health, Labour and Welfare.

With a similar mandate, in Argentina the National Advisory Committee on Agricultural Biosafety (CONABia) and in Brazil the National Technical Commission of Biosafety (CTNBio) are in charge of biosafety regulation and monitoring. In Argentina, CONABia under the Ministry of Agriculture is responsible for harmonizing policies relating to biosafety. In Brazil, CTNBio under the Ministry of the Science and Technology has regulatory duties related to biosafety. CTNBio is composed of 36 members, including representatives from the scientific community in human, animal, plant, and environmental sciences, and other national leaders. This commission meets monthly to certify the safety of laboratories and authorize experiments with genetically modified organisms, and also evaluates requests to commercially release genetically modified products.

BIOTECHNOLOGY IN AGRICULTURE . . . .

Ninety percent of the world's food is produced in North America, Europe, and Asia. More recently, Latin America has been increasing its contribution to the world food supply. For example, in 1979, Brazil produced about 39 million tons of grains. In 2000, that production increased to 84 million tons. The country was able to more than double its food production in a little less than 20 years. This increase is attributable to increased yields and the development of increased acres of farmland.

In the worldwide market, developing countries are small players in the global food supply. What role could biotechnology have for these countries? This science would allow increases in yield, improvement of nutritional quality, and reductions in agricultural production costs. This will be important, helping developing countries improve their economies by improving agriculture. One industry that is being strongly influenced by biotechnology is agro-chemical production. Worldwide, the agro-chemical industry is valued at about $20 billion (U.S.), annually. About $8 billion (U.S.) is related to agricultural chemicals used for control of diseases, insects, and weeds annually. In some crops, such as cotton, the cost of agro-chemicals accounts for nearly 40 percent of the total production cost. Biotechnology has made available pest-resistant cotton varieties that dramatically reduce the use of chemicals in crop management. This not only has a positive impact in reducing the costs for the farmer, but also benefits the environment by dramatically reducing the total amount of chemicals being used.

In the beginning of the 1990s, many people were skeptical of predictions that transgenic crop varieties would become a reality and would impact agriculture. The experience of the last six years, with the total area occupied with transgenic varieties reaching approximately 300 million acres, indicates that the initial forecasts about GMOs were wrong. In countries such as Argentina, more than 90 percent of the soybean acreage is planted with genetically modified soybean varieties. The continuous increase in farmland occupied with genetically modified varieties according to the 2001 census indicates their tremendous impact on agriculture worldwide. The acreage planted with transgenic soybean and cotton was, respectively, 63 percent and 64 percent of the total acreage in 2001. In the case of corn, the area planted with transgenics represented 24 percent of the total area, indicating stabilization in relation to the previous year. This stabilization is probably due to the reduction in the targeted insect pest population and to the smaller comparative advantage of the genetically modified varieties to normal corn hybrids when there is a low incidence of insects.

Transgenic Varieties

The first transgenic crops were field tested in the early 1980s. Currently, there have been more than 25,000 field tests worldwide, half of them in the United States and Canada. In Latin America, the largest number of field releases has occurred in Argentina. The commercialization of transgenic varieties began in the 1990s, with a tomato modified by Calgene. Today, transgenic varieties of various crop species are used in the United States, Canada, Mexico, Australia, France, Spain, China, Argentina, Uruguay, and other countries (see Figure 5-2). The plant species mainly have resistance to insects

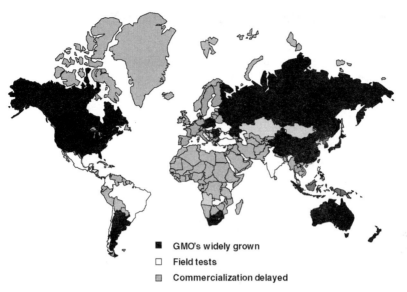

■ GMO's widely grown
□ Field tests
▨ Commercialization delayed

Figure 5-2
Global status of genetically modified crops in 2002.

and diseases, tolerance to herbicides, and improved nutritional quality, as in the case of canola, in which the fatty acid composition was altered to make products more suitable for heart patients.

Transgenic Animals

Additional difficulties in animal transformation have delayed progress in making genetically modified animals available in the market. The first transgenic animal marketed for use was the *oncomouse*, a mouse in which a gene for cancer had been introduced. The animal is being used in cancer therapy studies and the development of cancer-fighting drugs. The first transgenic animal for food production was the Atlantic salmon, which reached the U.S. market in late 2001. The transgenic salmon was modified to produce larger amounts of a growth hormone. The transgenic fish grow more quickly due to a nutrient conversion rate about 15 percent greater than that of conventional salmon. Other transgenic animals such as bovine, swine, and poultry are in the final steps of evaluation and they could be released in the market in the coming years.

CONCERNS ABOUT BIOTECHNOLOGY . . . .

The fast pace of biotechnology has raised many safety, ethical, and legal issues. One of the more sensitive issues came with the cloning of mammals from the somatic cells of adult animals. Dr. Ian Wilmut, at the Roslin Institute in Scotland, who worked with Dolly, the now famous cloned sheep, was the first to accomplish this. Today, there are clones of cattle, pig, monkeys, mice, and many other animals. News briefs covering the latest advancements and setbacks in this science are now commonplace. It is obvious that many questions remain unanswered.

Aside from the scientific questions, there exist social questions that need to be addressed. In the United States, there have been many public hearings in the federal legislature, and many states have subsequently banned human cloning. Additionally, some countries in Europe and elsewhere have raised concerns about the risks that cloned animals and transgenic crop varieties pose to the environment and human and animal health.

Food Safety

Companies releasing transgenic products must perform many tests to evaluate the safety of these products. For instance, data reported in the scientific literature suggests that the Bt protein of insect-resistant varieties and the modified EPSPS protein of herbicide-tolerant crop varieties in the market are safe for human and animal consumption. One such test, in vitro digestion, which simulates human digestion, has shown that there is no reason to suspect that these transgenic products present any risks to human health. This and other tests showed the products were suitable for release and use in human consumption. One product, Starlink corn, was licensed and released for use only in animal feed. The corn contains a protein that failed an in vitro digestion test, indicating the potential of becoming an allergen. Many corn-based foods were recalled when traces of Starlink corn were found in the products, but to date there have been no substantiated reports of health problems associated with the corn.

In the development of bioengineered products, biotechnology companies perform many in-house tests to assess biosafety prior to performing government-mandated tests. In some cases research on some traits is suspended when problems relating to human, animal, or environmental health are found.

Environmental Safety

Gene escape or gene flow, the exchange of genetic information among individuals or species, from transgenic crops is an often-discussed problem related to biotechnology. The development of super-weeds and loss of natural diversity is an important topic. This phenomenon can occur in three main ways:

1. When the transgenic plant itself becomes a weed
2. When the transgene is transferred by crossing from an adapted variety to a wild relative
3. When the transgene is asexually transmitted to other species and organisms.

Therefore, for gene flow to take place by means of sexual reproduction, certain conditions must be met:

1. The two parental individuals should be sexually compatible.
2. There should be an overlap in flowering time between the two parents.
3. An appropriate pollen vector should be present to transfer the pollen among the individuals.
4. The resulting progeny should be fertile and ecologically adapted to the local conditions where the parents are grown.

In the United States, soybeans and corn do not possess the biological attributes to escape and establish themselves as weeds. Corn is a cross-pollinated species with its pollen spread among plants by wind. Corn pollen has a low density and can travel short distances, varying according to wind patterns, humidity, and temperature. In general, cornfields are isolated from each other, especially for seed production, with distances of about 200 meters. The risk of gene flow of the soybean and corn to wild relatives in the United States is considered, for most scientists, small or nonexistent. In other countries, this risk might be significant.

In Mexico, scientists recently detected the presence of corn transgenes in wild corn relatives. Mexico is the center of origin of corn, and many wild corn relatives are found and even cultivated there. In this area, the risk of gene flow from cultivated corn to wild relatives is substantially larger. In fact, gene flow between the cultivated corn and its wild relatives has been reported for more than three decades in Mexico.

An article by Quist and Chapela (2001) raised concern about gene flow from genetically modified varieties to wild corn relatives in Sierra del Norte in the Province of Oaxaca in southern Mexico. However, on April 4, 2002, *Nature* sent ripples through the scientific community and the popular press by admitting it made a mistake. In an unprecedented action, it concluded in the journal's online version that "evidence available is not sufficient to justify publication" of the Quist and Chapela article. Reviewers agreed that Quist and Chapela's data were polymerase chain reaction (PCR) artifacts and did not support their conclusions.

In considering evolutionary concerns, it is important to mention the effects of transgenics on other organisms in the ecosystems. As mentioned, many transgenic crop varieties have been endowed with resistance to insect pests. The plant is able to produce certain proteins that are harmful when ingested by the insect but cause no harm to humans. One example is the Bt protein in corn varieties resistant to the European corn borer. With these transgenic plant varieties that have resistance to insects, concerns have emerged about how Bt varieties and other insecticide-producing plants will affect beneficial insects such as bees and ladybugs. The effect on these insects would have important implications on a broader scale. Scientific data suggests that the required lethal dose (LD_{50}) is much larger than the amounts insects will be exposed to in fields with these transgenic varieties. Although the safety of the Bt varieties for the monarch butterfly was initially questioned, subsequent research by different groups has indicated its safety for these insects (Tabashnik, 1994; Tang, 1996).

In the applications of biotechnology, one can find the importance of addressing the concerns of biosafety. The subject has many applications to the science as well as the social aspects of biology, but is focused on maintaining the safety and health of humans, animals, and the environment.

6 Cloning

In this chapter...

A *clone* can be defined as an individual or group of individuals that descend, through asexual reproduction, from a single individual. In other words, a clone is an exact copy of the original individual. Humans have practiced cloning of plant species for thousands of years. A leaf, a piece of stem, or root of a certain plant placed in a pot with soil or in a petri dish with tissue culture media can regenerate a new individual, genetically identical (clone) to the plant from which the leaf, stem, or root piece was taken. Today, cloning is a common agricultural practice used in many species that can easily reproduce asexually, such as sugar cane, banana, citrus, potato, strawberry, many grasses, roses, and many tree crops.

Cloning is based on two principles:

1. All cells of any organism contain the complete genetic makeup of the species.
2. Totipotence, the ability of one cell to differentiate and regenerate a completely new individual.

Although the regeneration of a complete plant from a somatic tissue (leaf, root, stem, etc.) is an ancient practice, it was only in the 1950s that biologists discovered the principles behind regeneration of whole individuals from a single cell. Unlike animal cells, most plant cells retain their potential to express any of their genes and therefore are able to repeat the developmental processes involved in regenerating complete individuals. Cloning of plants offers the possibility of developing millions of individuals exactly identical to the original source of the regenerated cells. This is a common method of reproduction in asexual plant species.

Most animal cells do not have that same capability of naturally regenerating a complete individual from a cell. In animals, this potential is lost during cell specialization. A specific class of cells called *stem cells* is the only cell type known to retain their totipotence. Stem cells can be found in marrow tissue, fat tissue, and developing embryos. These types of cells have been the focus of animal cloning efforts. In animals, cloning can be accomplished using the technique called *nuclear transplant*. The technique has been used for many years in animal cloning using embryonic cells for amphibians such as toads. Animal embryonic cells maintain their totipotence after the first few cellular divisions. As the embryo continues its development, the cells lose their ability to differentiate into other cells and, consequently, the capability for complete regeneration ceases quickly. Contrary to the relative ease of nuclear cell transfer in amphibians, this process is much more complex in mammals. Although cloning of toads was accomplished for the first time in 1952, cloning of mice using the same technique was not accomplished until 1977.

Cloning using nuclear transfer involves the manipulation of two cells. The recipient cell is usually a nonfertilized egg from a female taken soon after ovulation. Harvesting of these eggs is done by laparoscopy or by transvaginal suction. The donor cell, which is the one providing the genetic material for regenerating the clone, is collected from the individual to be copied. Any somatic cell could be used for the purpose, including cells from the skin, mammary glands, or mucous membranes. Under a microscope, the recipient cell (egg) is held, by suction, at the end of a pipette. With an extremely fine micropipette, the chromosomes are removed. At this point the nucleus from the donor cell is then fused with the recipient egg previously deprived of its chromosomes. Some of the cells, if implanted into the uterus of a surrogate mother, start developing into an embryo and eventually a fetus. The procedure involves the removal or destruction of the chromosomes from the recipient egg cell, and the subsequent introduction of the chromosomes from the donor cell. The egg, with the newly introduced genetic material, begins the developmental process in the uterus of a surrogate mother to form a complete individual, genetically identical to the donor that supplied the nucleus. This technique has been used with success for cloning sheep, cattle, mice, monkeys, and other mammals.

The first cloned mammal from somatic cells of an adult donor was the sheep Dolly, born in February 1997. Dolly was cloned using mammary cells from an adult sheep. This widely covered event occurred at the Roslin Institute in Scotland, and the lead scientist was Dr. Ian Wilmut (Figure 6-1).

The main players in animal cloning are Geron Corporation, Advanced Cell Technology (ACT), PPL Therapeutics, and Infigen. Geron (*http://www.geron.com*) is a biopharmaceutical company focused on developing and commercializing therapeutic and diagnostic products for applications in oncology and regenerative medicine, and research tools for drug discovery. It is researching embryonic

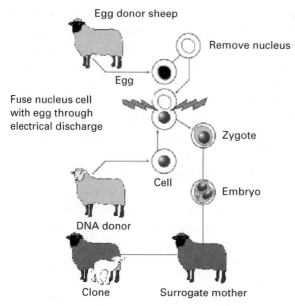

Figure 6-1
Outline of the cloning procedure used for creating
Dolly in 1997.

stem cells for treating disease. ACT (*http://www.advancedcell.com*) was
the first to clone a human embryo for therapeutic purposes. ACT is
involved in nuclear transfer for human therapeutics and animal
cloning. PPL Therapeutics (*http://www.ppl-therapeutics.com*) applies
transgenic technology to the production of human proteins for therapeutic and nutritional use. PPL Therapeutics, in partnership with the
Roslin Institute, cloned the first mammal from an adult cell, Dolly.
Infigen (*http://www.infigen.com*) commercializes applications of
cloning technologies and genetic testing in the cattle breeding, pharmaceutical, and nutraceutical fields.

TECHNICAL HURDLES IN CLONING . . . .

There are two major problems or limitations found in the cloning of
mammals. First, following the introduction of the donor's nucleus
into the egg, it must be reimplanted into a gestating surrogate

mother. Most of the implanted eggs abort, forcing scientists to perform several implantations, in the hopes that at least one of the females will have normal gestation. In the case of amphibians (e.g., toads), the development of the embryo occurs outside of the adult's body, thereby facilitating the development of the fetus.

The second major challenge in animal cloning is the size of the fetus at birth. Most of the surrogate mothers have to deliver via Cesarean section. This is especially true in bovines, as the clones tend to be about twice as big as normal newborn calves. The large size of the fetus during gestation can represent a substantial risk for the surrogate mother. Additionally, clones tend to have a high incidence of birth defects, and many clones die in the first hours following birth. Common abnormalities observed in cloned animals include failures of the kidney, heart, circulatory system, liver, and lungs. In addition, the placenta of the surrogate mother does not always function properly during gestation.

The causes of the high abortion rate and abnormalities in clones are still not completely understood, but it is suspected that they are at least partially the result of the complexity of the genetic reprogramming that takes place in the genes from the donor that are inserted into the egg. If a gene is expressed inadequately or it is not expressed at a critical point in development, the result can be a developmental defect. Genetic reprogramming involves the regulation of thousands of genes in a systematic and orderly way. Any asynchrony in the expression of the genes can contribute to defects in the fetus or even result in abortion. Additionally, when cloning is done with nuclei from somatic cells, they bear any preexisting mutations that might have occurred after the cells had differentiated into specialized cells. These mutations would have otherwise been screened out in gametogenesis.

With the current knowledge and technology, mammalian cloning is still a highly unsafe and inefficient procedure. The expectation is that, as new knowledge is generated from more experience, the main limitations in cloning will be at least partially solved. This science is continuing to make progress worldwide, even in developing counties. For example, in Brazil, Embrapa-Cenargen recently pioneered the cloning of the first bovine calf from somatic cells, born in March 2001 (see Figure 6-2).

Figure 6-2
Vitoria, the first calf cloned in Brazil by
Embrapa-Cenargen.

THERAPEUTIC CLONING . . . .

Most people and the scientific media were surprised with the publication of the article "The First Human Cloned Embryo" in the magazine *Scientific American* in November 2001. Even the most optimistic
followers believed that experiences with human cloning would not
produce results so early. ACT, a small biotechnology company in
Massachusetts, was the first to accomplish the cloning of human cells
for therapeutic use. Dr. Michael West, the company's chief executive
officer, emphatically stated that his company's objective was research
in cloning for exclusively therapeutic treatments and not for reproductive or human cloning purposes. Nevertheless, the public had a
strong reaction to this news.

Those who favor the use of embryonic stem cells tend to see the
potential to cure genetic problems, and they emphasize the hope of a
cure and improved lives for patients with fatal genetic diseases. Opponents recognize that human life begins at conception, and they believe that the price of a cure should not result in the taking of another
life, as harvesting embryonic stem cells for this research results in the
destruction of embryos. Therefore, embryonic stem cells can be seen
as a matter of life by those who can benefit from this technology, or
as a matter of death by those who do not agree with the sacrifice of
embryos for the production of stem cells.

This is not an easy debate. Imagine a case in which the only hope of cure for a young mother with two small children is the use of embryonic stem cell therapy. Even if this mother's dramatic situation might suggest that it would be ethical to sacrifice a mass of frozen cells stored in liquid nitrogen to obtain the needed stem cells for the therapy, the point that deserves to be addressed is this: Who would have the right to sacrifice a defenseless life (embryo) to save another (adult individual)?

The use of stem cells from bone marrow, umbilical cord, and other parts of the adult human body has not generated as much controversy. The potential benefits from stem cell therapy have been widely discussed. However, the use of embryonic stem cells has raised heated debates in public and scientific arenas. These cells are usually harvested from spare embryos generated through in vitro fertilization that have not been implanted in prospective mothers. Even if the scientist that uses stem cells were not responsible for producing them, he or she would be aiding in this process by creating a demand that results in the destruction of embryos, being an accomplice in the process. This is the same rationale used by the governments that burn ivory confiscated from smugglers, as well as the refusal of the scientific community to use the knowledge generated by the Nazis in the horrific human experiments conducted at the concentration camps during World War II.

This and many other recent discoveries in biotechnology have been occupying the world media. Although the scientific bases for cloning are easy to understand, the greater challenge for society is to address its ethical issues.

The lack of ethical references and the speed of development of new knowledge have exposed the society's lack of readiness to address current ethical issues. Sometimes society fears a technology with great potential benefits; other times it is apathetic about technology with proven negative impacts. Individualism and relativist morale, ideals in fashion in this postmodern society, are fertile ground for justifiable mistakes. These ideologies emphasize that nobody should deny anything to himself or herself that is good unless it is especially harmful to his or her neighbor. The ethical boundaries of society reflect the moral principles that it possesses. Society is dynamic and so are its ethical values. This doesn't mean, however, that the principles within society should develop in a liberal way.

Humans were created with intelligence and this allows them to develop new technologies and expand science. Along with this intelligence they have the freedom to choose between good and bad.

Why should one not be in favor of the evolution of the human race? What are the limits of what is morally acceptable? Any answer that deserves consideration should address the dilemmas of society in light of its principles, morals, and religious beliefs. These are some of the challenges society must deal with.

For more information on cloning refer to the following Web sites:

- National Academy of Sciences:
  http://www4.nationalacademies.org/nas/nashome.nsf
- *New Scientist* Magazine:
  http://www.newscientist.com/hottopics/cloning

HUMAN CLONES . . . .

After the cloning experience of Dolly the sheep, human cloning is theoretically and technically possible. The procedure would consist of taking an egg, removing its chromosomes, and then fusing it with a somatic cell from the individual to be cloned. Some believe that it is inevitable that some scientist will try to clone humans, if it is not already occurring. There seems to be a consensus that within a few years the news of the birth of the first human clone will be the major headline in the media. Scientists in South Korea reported success in creating a cloned human embryo, but it was destroyed instead of being implanted in a surrogate mother. Even if the first human clone is decades from birth, the idea that scientists are secretly trying to do it is a real possibility.

Scientists with an economic interest in this science have been expressing their viewpoint that it would be ethically acceptable to clone human beings. They argue that an embryo up to 10 days after fertilization cannot be considered a life because development of the brain begins at about 14 days after fertilization. It would be interesting to know how those scientists define the ethical limits in relation to their objectives.

It has been assumed by some that human cloning serves only the interests of the narcissists or neo-Nazis, those who would like to create the perfect race. In fact, several scenarios have been created that justify cloning of the *Homo sapiens* "animal." Some of those scenarios can seem extremely appealing, but an ethical analysis of the dilemmas that clones, their relatives, and society would face during their life indicates that cloning of the most intelligent and rational of the animals is not politically, socially, or religiously acceptable.

Some of the following scenarios show the complexity of the subject:

- Consider the situation of a homosexual man who feels frustrated with his incapacity to bear children and wants to be cloned.
- Consider the couple that wants to have a baby, but the husband is sterile. Assuming that cloning is an alternative, the couple could decide to clone the husband, and the wife could contribute as a surrogate mother. Would the child's responses to education differ though he is genetically identical to his father? Would he have the same tastes and preferences as the husband? What if a divorce occurs? How would the mother see her son, who is a copy of the man from whom she is divorced? Would the father have the right to custody of the child because he is genetically related to his father?
- In another scenario, where a woman gives birth to her own clone, would she be her child's mother or twin sister with a different age?

Obviously, society changes over time. In vitro fertilization was illegal in many countries until about 20 years ago, and the idea of heart transplants was considered immoral in the past. Public opinion on human cloning will probably change in the next few years, but cloning will likely be banned globally before the birth of the first human clone. It would be a terrible mistake to wait until the birth of a baby with genetic defects before that decision is reached. Current experience with animals shows that this technology has too many technical and ethical problems to justify experimentation in humans.

Ethicists are concerned that clones would be considered inferior to human beings, and they would be subject to the limitations and expectations of the knowledge of the copied person. These expectations could be false, as both genetic factors and the environment determine personality. For example, a clone of an extroverted person could be more introverted, depending on his or her upbringing. Clones of athletes, artists, scientists, and politicians could choose different professional careers based on opportunities and the environment in which they are raised.

Predicting the future of human cloning is not an easy task. History shows that society is dynamic, that ethical values change, and moral principles distort over time. In other words, only time will tell. The challenge for bioethicists is to keep science progressing while maintaining the sanctity of life. Additional ethical arguments related to human cloning are presented in Chapter 14, "Bioethics."

It is a mistake to think that genetically identical means identical individuals. In the 1978 movie *The Boys of Brazil*, based on Ira Levin's bestseller, a scientist conspires after World War II to clone Hitler, with the objective of raising a new generation of Nazi leaders. The film shows that without intense indoctrination, the clones can be influenced to pursue other activities than becoming dictators.

Scientific Reasons for Not Cloning Humans

There is a series of scientific reasons for not cloning human beings. Although many scientists and most of the public share this point of view, it is feared that personal ambition of unscrupulous scientists would make them blind to the scientific reasons for not cloning man. The success in animal cloning is evidence that this technology might be ready to justify its application to humans. In 2000, Dr. Panayiotis Zavos, an Israeli specialist in in vitro fertilization, and Dr. Severino Antinori, an Italian specialist in reproductive physiology, announced their intention to clone humans. In April 2002, Antinori claimed that he had two women carrying cloned babies.

After the birth of Dolly and the successful cloning of mice, cattle, monkeys, goats, and pigs, it is evident that cloning is not a completely safe procedure. Cloning of mammals is considered highly inefficient, and this is unlikely to change in the foreseeable future. Many cloning experiments have resulted in developmental flaws either during gestation or in the neonatal period. Even in the best cases, only a small percentage of cloned embryos survive to birth and, of those, many die shortly after birth. There is no reason to believe this would be any different with humans. This means that to achieve the successful generation of a human clone, many others will have been sacrificed in the developmental phases.

The few animal clones that have survived and been born show abnormal size, a phenomenon called *increased offspring syndrome*. It is believed that incorrect functioning of the placenta is one of the main causes of embryonic death. The suspected causes of newborn death are respiratory and circulatory problems. Some seemingly healthy survivors might possess immune system dysfunction or kidney and brain malformation. Those problems have been detected in practically all species in which cloning has been accomplished. Therefore, if an attempt to clone a human is made, the concern is not just with the embryos, but also with those that will live to be abnormal children and adults.

The abnormalities in the fetuses and in those few clones that are born alive cannot be easily traced to the nucleus of the donor. The most probable explanations are flaws in the genetic reprogramming or timing and expression of the correct developmental genes. Normal development depends on a necessary sequence of changes in the configuration of DNA and proteins coded by developmental genes. Those developmental changes control the specific genetic expression in the specialized tissues.

Genetic reprogramming of the entire genome is a natural process that happens during spermatogenesis and oogenesis, which can span over months and years in humans. During cloning, reprogramming of the donor's DNA must be done within minutes or, at the most, in a few short hours, during the period of time that nuclear transfer is completed and cell division begins to form the zygote.

Prenatal mortality in clones can occur due to inadequate reprogramming that results in improper gene expression. Some surviving clones have subtle genetic defects that, over time, result in life-threatening conditions. There is no information on genetic regulation in clones, but some evidence seems to indicate errors in gene expression in cloned animals. The expression of marked genes is significantly altered when embryos are cultivated in vitro before they are implanted in the uterus, indicating that even a minimal disturbance of the embryo's environment can have profound effects on gene regulation during development.

All the current evidence now suggests that the experiments on human cloning announced by Zavos and Antinori will have the same failure rates and occurrences of abnormalities that have been detected in animal cloning. Zavos tried to calm the public, informing them their research would use genetically perfect embryos to be implanted as a quality control. However, the public perception of reproductive biotechnology will be seriously damaged if the research fails and defective babies are born from human cloning experimentation. This would likely negatively affect other areas of research, such as the advancements being made with stem cells.

The National Bioethical Advisory Commission in the United States reached the following conclusion six years ago: "At the present, the use of cloning to generate a child would be a premature experiment, and would expose the fetus and the child in development to unacceptable risks." All the data gathered since seems to reinforce this point of view. In many countries, it is unlawful to perform research with human reproductive cells, thereby forbidding embryonic cloning.

Other ethical considerations of human reproductive biotechnology are discussed in Chapter 14.

7 Gene Therapy

In this chapter...

Gene therapy has become an increasingly important topic in science-related news. The basic concept of gene therapy is to introduce a gene with the capacity to cure or prevent the progression of a disease. Gene therapy introduces a normal, functional copy of a gene into a cell in which that gene is defective. Cells, tissue, or even whole individuals (when germ-line cell therapy becomes available) modified by gene therapy are considered to be transgenic or genetically modified. Gene therapy could eventually target the correction of genetic defects, eliminate cancerous cells, prevent cardiovascular diseases, block neurological disorders, and even eliminate infectious pathogens. However, gene therapy should be distinguished from the use of genomics to discover new drugs and diagnosis techniques, although the two are related in some respects. The two main types of gene therapy are somatic cell gene therapy and reproductive or germ-line gene therapy. This chapter also discusses therapeutic cloning, which involves stem cell manipulation for tissue and organ production.

Germ-line cell therapy involves the introduction of corrective genes into reproductive cells (sperm and eggs) or zygotes, with the objective of creating a beneficial genetic change that is transmitted to the offspring. When genes are introduced in a reproductive cell, descendant cells can inherit the genes.

Gene therapy of somatic cells, those not directly related to reproduction, results in changes that are not transmitted to offspring. An example of gene therapy in somatic cells is the introduction of genes in an organ or tissue to induce the production of an enzyme. This alteration does not affect the individual's genetic makeup as a whole and it is not transmitted to its descendants. With somatic cell gene therapy, a disabled organ is better able to function normally. This technology has many applications to human health. One variant of somatic cell gene therapy is DNA vaccines, which allow cells of the immune system to fight certain diseases in a method similar to conventional vaccines.

Stem cell therapy involves the use of *pluripotent cells,* or cells that can differentiate into any other cell type. Stem cells are found in developing embryos and in some tissues of adult individuals. This therapy is similar to a conventional transplant, with the objective of regenerating or repairing a damaged organ or tissue. The procedure has a reduced probability of rejection because it uses the individual's own cells. For instance, stem cells differentiated into nerve cells could be used by patients suffering from paralysis, with the goal of helping them recovering movement; or in cases of heart stroke, muscle cells might be used to rejuvenate the cardiac muscles. Furthermore, the future may bring the growth of stem cells from an individual's body to produce certain tissues or organs in vitro. Stem cell research could eventually blend gene therapy with genetic engineering to create healthy stem cells that can be used to generate healthy organs and tissue.

A fundamental requirement for gene therapy is the correct identification of genes coding for diseases. This can be accomplished at a spectacular speed with the information from the Human Genome Project. Scientific magazines have been announcing, with great frequency, the discovery of genes responsible for several medical conditions, from Alzheimer's disease to baldness. The knowledge of the genes involved in these traits allows unequivocal diagnosis of the disease in the patient, an essential step before treatment can be initiated for the genetic disease. Biotechnology is contributing to the development of the needed genetic tests for detection of defective genes.

The most complex phase in gene therapy is the development of mechanisms to deliver the therapeutic genes to the target organ in an accurate, controlled, and effective way. That step has been developing more slowly and is currently the most limiting factor for gene therapy.

GENETIC DEFECTS . . . .

Each human being carries normal as well as some defective genes. Usually, the individual does not become aware of the presence of a defective gene until a disease associated with the gene is manifested in him or her or in a relative. More than 4,000 medical disorders caused by defective genes have been identified, each with varying degrees of seriousness. About 10 percent of the human population will evidence, sooner or later, some type of disorder. Although genes are responsible for predisposition to disease, the environment, diet, and lifestyle can affect the onset of the illness.

An example of a genetic disease is cystic fibrosis, which frequently becomes evident in the first years of life for the child carrying the defective gene. The mutant gene causes the development of cysts and fibrous tissue in the patient's pancreas and the production of thick and viscous lung mucous. The mucous makes breathing very difficult and, in many cases, is fatal. On average, in Western countries, about 1 child in 2,500 has the disease. If the child receives two defective recessive alleles of the gene named CF (one from each parent), he or she will develop the disease. Patients with cystic fibrosis can reduce the symptoms of the disease with drugs developed through genetic engineering. A cure for cystic fibrosis may come through gene therapy. One possibility is a genetically engineered virus, carrying the corrective gene, which after being introduced into the patient's lung cells would allow the lungs to function properly. The introduced gene would allow the lung cells to produce a protein that eliminates the mucus.

Most people do not manifest genetic diseases because, most of the time, they are carriers of just a single defective copy of the CF gene. As most of the defective genes are recessive, meaning two copies are needed for expression of the disease, most people do not have the disease. This is the reason for the larger incidence of genetic diseases in children from related parents.

If the defective gene, however, is dominant, the disease is expressed in any people that carry the defective gene. Huntington's Disease, a disorder of the nervous system that usually occurs after the age of 45, is an example of a genetic disease caused by a dominant gene.

Having a defective gene does not make disease development a certainty. Besides the large effect from genetics, the environment is also important to the onset of many illnesses. Diseases such as heart disease do have a genetic component, but are largely dependent on diet and lifestyle. Some genetic diseases also have benefits. A classic example of a genetic disease that has a beneficial effect on human survival is sickle cell anemia. There exists in the human population a defective β-hemoglobin gene and individuals carrying two copies of the defective gene develop sickle cell anemia, a blood problem caused by defective hemoglobin and consequently misshapen red blood cells. The genetic mutation in the defective allele of this disease is a single nucleotide change, from an A in normal genes to a T in the mutant. This single nucleotide mutation results in a mutant β-hemoglobin that possesses the amino acid valine instead of

Table 7-1
Medical Conditions for Which Gene Therapy Is Being Studied

ADA deficiency	Hemophilia
AIDS	Liver cancer
Asthma	Lung cancer
Brain tumor	Melanoma
Breast cancer	Muscular dystrophy
Colon cancer	Neurodegenerative conditions
Diabetes	Ovarian cancer
Heart diseases	Prostate cancer

glutamine. The mutant β-hemoglobin has less affinity to oxygen, becoming a poor oxygen transporter in the blood. However, carriers of a single copy of the defective allele do not have the disease, and they are also resistant to malaria. There is an obvious advantage of carrying a single allele of the defective hemoglobin gene, especially in regions where malaria is endemic, as in tropical regions of Africa.

The first case of gene therapy occurred in 1990, at the NIH in Bethesda, Maryland. On that occasion, a four-year-old patient with a severe immune system deficiency (adenosine deaminase enzyme [ADA] deficiency or bubble-boy disease) received an infusion of white blood cells that had been genetically modified to contain the gene that was absent in his genome. Since then, gene therapy has been studied and experimentally tested for several medical conditions.

Diseases caused by the absence of an enzyme or the presence of an inactive enzyme are potential targets for gene therapy. Cystic fibrosis, ADA deficiency, and many other genetic diseases are among the candidates for gene therapy. Table 7-1 lists other diseases for which gene therapy is being considered.

VECTORS FOR GENE DELIVERY . . . .

Appropriate methods to deliver DNA used in gene therapy are vital, as the targeted tissues must properly receive the appropriate genes. Gene therapy can be carried out using naked DNA delivered directly into the target cells. However, this procedure of introducing isolated DNA molecules has a very low efficiency rate. To increase the efficiency of DNA uptake by the target cells, special vectors have been engineered for gene transfer. *Vectors* are plasmids or viruses that are used to move recombinant DNA from one cell to another. A retrovirus is a special class of RNA viruses that can insert its nucleic acid into host cells. The viruses possess a gene for production of the reverse transcriptase, an enzyme that transcribes RNA in DNA in the host cell. Adenovirus, retrotransposons, and liposomes are other vectors used for gene transfer in gene therapy. They are all able to transfer and integrate genes into new cells. Retroviruses used in gene therapy are engineered so that any genes that are harmful to man are removed. Corrective genes are then added to replace the removed genes, and the new, modified retrovirus is then introduced into the patient.

One of the challenges for vectors is to survive the patient's immune system so they can transfer the corrective genes from their genome into the patient's cells. In general, the immune system of the human body contains molecules that immobilize viruses or other microorganisms that could infect the organism. Viruses that escape the immune system need to penetrate the cellular membrane, an additional barrier to infection. Finally, the infecting retrovirus must integrate its genome with that of the host, thereby moving the corrective genes into the genome of the infected cell. This integration happens in a random manner. It should occur in an area of DNA that is not essential to the host genome, or a risk of other complications might occur. Furthermore, the introduced gene must be transcribed and expressed for the production of the correct enzyme. With all these processes at the molecular level, gene therapy becomes a very complex procedure.

Another promising strategy, which has been used for the introduction of therapeutic genes in lung cancer treatment, is the direct injection of the corrective genes into the target area. Using this strategy, scientists have injected a drug containing the normal version of the gene p53, which suppresses cell tumor growth, directly into the patient's cancerous tumor. This technique bypasses the immune system reaction to the invading vector, a problem frequently associated with gene therapy. Many scientists believe that as gene therapy develops, it will be possible in the near future to easily introduce genes into patients through intramuscular injection, especially for cases of anemia, hemophilia, diabetes, and other diseases related to the circulatory system.

GENE THERAPY RISKS . . . .

The first death associated with gene therapy occurred on September 18, 1999, at the University of Pennsylvania. Jesse Gelsinger was participating in a clinical trial, a biomedical experiment for evaluation of safety and efficiency of a therapy for a disease. Gelsinger, who was 18 years old at the time of the treatment, had a deficiency of ornithine transcarboamylase, an important enzyme in the metabolism of ammonia. Patients with this rare metabolic disorder must maintain a low-protein diet and take a series of medicines to avoid ammonia poisoning in the blood stream. The gene therapy Gelsinger took triggered a chain reaction in his immune system, resulting in hepatic and respiratory failure, and consequently, his death four days after being treated.

Since Gelsinger's death, the University of Pennsylvania has been reevaluating all procedures involved in the vector engineering and in the administration of the therapy. No flaw has been found that would explain such an extreme reaction by his immune defense system. Ever since, the public and the FDA, the agency responsible for oversight of clinical trials in the United States, have been more skeptical and doubtful about whether current scientific knowledge is enough to justify further investigations with humans. The credibility of gene therapy was seriously damaged, resulting in a temporary moratorium on human clinical trials.

Another challenge to gene therapy has been its ephemeral benefits to patients. This has been observed in several clinical trials with cystic fibrosis and ADA deficiency patients, whose cure faded after a few months of therapy, and was followed by a return of the disease symptoms. A possible explanation for that is that the genetically modified somatic cells (see Figure 7-1) decreased in amount. Because they are already differentiated and possess only a limited capability to multiply, it is expected that after they are gone, the treated organ could become diseased again.

Figure 7-1
Isolation of corrective genes for use in
gene therapy.

DNA VACCINES . . . .

A variation of gene therapy with somatic cells is the introduction of genes (naked DNA), with the objective of triggering the immune system to produce antibodies for certain infectious diseases, cancer, or some autoimmune diseases. Therefore, the objective is not repair of a defective gene in the individual's genome. Those genes can be introduced via intramuscular injections, inhalation, or oral ingestion. Cells that take up the gene in their genome can express the protein that stimulates the immune system to act against the disease.

The greatest challenge in this procedure is the transient effect of gene expression, because the modified cells can go through only a limited number of divisions before dying. Another challenge is the low efficiency of gene incorporation and expression in the target cells. Although in some cases the temporary gene expression is enough to trigger an effective immune response, most cases require a more lasting gene expression.

GERM-LINE CELL THERAPY . . . .

The main advantages of germ-line cell gene therapy are the following:

1. It offers the possibility for a true cure of several diseases and it is not only a temporary solution.

2. It might be the only way to treat some genetic diseases.

3. The benefits would be extended for several generations, be-
    cause genetic defects are eliminated in the individual's
    genome and, consequently, the benefits would be passed to
    his or her offspring.

Some of the arguments presented against germ-line cell gene therapy are the following:

1. It involves many steps that are poorly understood, and the
    long-term results cannot be estimated.

2. It would open the door for genetic modifications in human
    traits with profound social and ethical implications.

3. It is very expensive and it would not benefit the common
    citizen.

4. The extension of the cure to a person's offspring would be
    possible only if the defective gene was directly modified,
    but probably not if a new gene was added to another part of
    the genome.

STEM CELL THERAPY . . . .

Stem cell therapy or therapeutic cloning does not involve gene therapy itself. However, in the future it might be used in conjunction
with gene therapy for regeneration of tissue and organs after they
have been treated with corrective genes. Visually, stem cells are not
distinguishable from any other cells of the human body. Under a
common microscope (magnification 20 to 40 times), those cells can
only be observed using special dyes. Visually there is no significant
difference in such cells. The real differences exist at the DNA level,
where gene expression is amendable to signals influencing protein
expression. The cells can differentiate into any of the 220 cell types of
the human body (e.g., kidneys, heart, liver, skin, or retina), a phenomenon called *pluripotency*. At birth, stem cells can be harvested
from an individual's bone marrow, fat tissue, and the umbilical cord.
Embryonic stem cells are harvested from embryos up to a few days
after fertilization.

Another characteristic of stem cells is their capability to grow indefinitely. Whereas the remaining body cells have a biological programming that limits the number of cell divisions they can go through before dying, stem cells can be maintained indefinitely in a petri dish with nutritive media.

Stem cell therapy provides hope for a cure for patients of incurable afflictions such as Parkinson's disease and Alzheimer's disease, and also for people suffering from paralysis resulting from spinal cord injuries.

At first, some opponents speculated that stem cells would be used in nurseries to produce organs such as livers, hearts, and virtually any other body part. However, most organs possess complex structures with ducts and valves, making it impossible to produce them outside of the organism. Stem cells have opened a new avenue for disease treatment. For example, the injection of stem cells into the liver of a patient with cirrhosis or hepatitis could result in new tissue capable of performing its role. Stem cell therapy also has great potential to cure rheumatoid arthritis and some heart diseases. Recent research has found that spine-injured mice suffering from paralysis were able to move their legs following an injection of stem cells.

Some people believe that if human stem cells are as versatile as those of mice, they might be the long sought after fountain of youth. The combination of stem cells with gene therapy might allow rebuilding of new body parts to substitute for old and defective ones. Right now, different procedures are being tested for curing ADA deficiency. Somatic cell gene therapies have the limitation of lasting for only a few months, which in turn requires repeated applications. With the use of stem cells to regenerate healthy bone marrow cells, a permanent cure is expected, as healthy cells have the capability to grow and divide continuously.

Embryonic stem cells, from embryos about four days old, have been at the center of a heated debate due to ethical issues. The main disagreement is whether or not a four-day-old embryo is already a human life. When would an embryo or a fetus reach the status of life? Those that support the use of embryonic stem cells would argue that human life would not begin until about the 14th day after the fertilization, whereas the opposition argues that life begins at conception (i.e., at the moment of the fertilization of the egg by the sperm). For many, the destruction of embryos for the purpose of treating another human being is wrong. Recently, in the United States, the Bush administration broadened the definition of a child eligible for coverage under the Children's Health Insurance Program by classifying a developing fetus as an "unborn child." Many activists are arguing that the Bush administration's proposal demonstrates its commitment to the strategy of undermining a woman's right to choose abortion by ascribing legal rights to embryos. This subject is addressed further in Chapter 14, "Bioethics."

FINAL CONSIDERATIONS . . . .

Although the idea of gene therapy has been around for only 20 years, the technique has been drawing a great deal of interest and curiosity through the world. The first trials generated great expectations within the scientific community. Although there have been several disappointments, many believe that it is just a matter of time before the technical and scientific details are mastered and the procedures become routine. This research is being advanced worldwide. In fact, Alain Fischer, a medical doctor in Paris, France, reported the complete cure of two children who had a rare immune deficiency condition.

Another promising result from stem cell research has been reported in type-B hemophilia patients at the Children's Hospital in Philadelphia and at Stanford University, where patients treated with gene therapy presented a reduction in the period for blood coagulation. ADA deficiency, a disease caused by a defective gene for the ADA enzyme present on human chromosome 20 has been a focus for gene therapy in many institutions. In one of the cases, several patients treated with the corrective gene were able to reconstitute their immune systems and are living normal lives out of the isolated bubbles that are needed to maintain an environment free from microbes. The patients started to produce a correct ADA enzyme after receiving the gene therapy.

The potential use of this therapy to cure other more complicated diseases, such as cancer and coronary diseases, also seems promising. Gene therapy is still in its infancy, but it is believed that as it matures, it will become an effective treatment for the myriad of genetic diseases that affect humanity.

8   Pharmacogenomics

In this chapter....

An old question that has long puzzled medical researchers is the variety of responses to pharmaceuticals by different people. The human genome project has helped to launch the new science of pharmacogenomics, which studies the association between an individual's genetic constitution and his or her response to medicines. Pharmacogenomics appeared from the observation that some drugs work well in certain people, but not in others. The study of the genetic basis of the differential response to therapies in different patients is allowing for the development of more efficient and safer medicines. The new generation of pharmaceuticals coming to the market is proof of that (Figure 8-1).

Society received with enthusiasm the news of the arrival of Gleevec, recognizing it to be a hallmark in the race for a cure for cancer. Gleevec was developed for a rare type of leukemia caused by a genetic mutation. When scientists clinically tested Gleevec, the results were impressive. In almost all cases, patients' blood tests returned to normal levels; in half of the cases, the cancerous cells disappeared or were reduced. The results were so promising that the FDA granted Gleevec approval for public use in record time. Gleevec is an example of a drug developed by pharmacogenomics. Today,

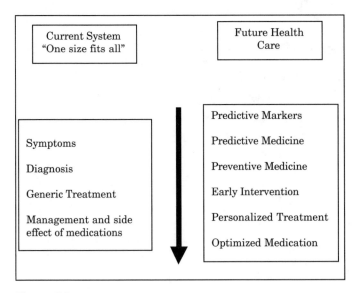

Figure 8-1
Health care in the 21st century.

Table 8-1
Pharmaceuticals in Cancer Clinical Trials

Drug	Laboratory	Target	Status
IMC-C225	ImClone Systems	Block growth factor receptor in cancerous cells	Clinical trials in colon, pancreas, brain, and lung cancers
GVAX	Cell Genesys	Vaccine that triggers the patient's immune system to attack cancerous cells	Clinical trials for prostate, lung, pancreas, and melanoma cancers
ABT-627	Abbott	Block receptor in cancerous cells	Clinical trials in prostate cancer
SU5416	Pharmacia	Block signal factor responsible for growth and diffusion of cancer cells	Clinical trials in colon and breast cancers

several other pharmaceuticals developed from pharmacogenomics are in the final phases of clinical trials (Table 8-1).

Pharmacogenomics and genetic engineering are designing drugs to combat the causes of diseases instead of their symptoms. This in itself is a revolution in medical therapy. To develop new medicines, the pharmaceutical industry is establishing partnerships with biotechnology companies. Until recently, the pharmaceutical industry spent, on average, from 12 to 15 years and about $500 million to develop a new medicine. During that period, scientists studied and tested hundreds of substances to identify a small number of drugs that were considered safe and promising for final evaluation in clinical trials. Still, some of the medicines do not work well in all patients, which is understandable because each person possesses a different genetic makeup (see Figure 8-2). Additionally, the side effects of a certain drug are known only after it is prescribed for a large number of patients. For instance, the FDA removed the medicine Rezulin from the market in 2000. Rezulin, a prescription drug for diabetics, was connected to 63 deaths that year.

In the next 365 days, about 200,000 women (and 1,500 men) will be diagnosed with breast cancer, up from 100,000 two decades ago. At

Figure 8-2
Genetic tests: The analysis of DNA from a blood drop can reveal intolerance to a certain medicine.
Source: Courtesy of ICN Pharmaceuticals, Inc.

first it might seem that the incidence of cancer is up, but the high numbers are in part a result of better detection techniques.

Biotechnology is facilitating the development of new medicines, making them faster, cheaper, safer, and more efficient. Instead of just studying the new drugs in clinical trials, scientists are identifying the genetic cause of the diseases. By simulation, they are able to design and study the action of new pharmaceuticals.

The partnership between the established pharmaceutical industry with more than a half century of experience and the newer biotechnology companies, in general less than 10 years old, is very promising. Millennium, Orchid BioScience, and many other companies that are pioneering the revolution in pharmacogenomics were established in recent years (Table 8-2).

Table 8-2
Partnerships for Pharmaceutical Development Through Molecular Genetics

| Industry | | Partnership | |
Pharmaceutical	Biotechnology	Date	Target
Aventis	Millennium	6/2000	Inflammatory diseases
Bristol-Myers	Entelos	12/2000	Obesity
Bayer	CuraGen	1/2001	Diabetes and obesity
AstraZeneca	Orchid BioScience	2/2001	Cancer, heart diseases
Abbott	Millennium	3/2001	Diabetes and obesity
Hoffman-La Roche	DeCode Genetics	10/2001	Various

Genomics is still in its infancy, and the human genome still holds many secrets that are just beginning to be revealed. For instance, it was only after the first draft of the human genome sequencing, in February 2001, that scientists realized that humans possess only about 30,000 genes, much smaller than previous estimates of 70,000 to 140,000 genes. It is therefore difficult to estimate the benefits from pharmacogenomics with so much still unknown. However, the clinical success of Gleevec seems to indicate a bright future for pharmacogenomics.

PHARMACEUTICALS AND SIDE EFFECTS . . . .

Many pharmaceuticals currently in use have drastic side effects on the body. An example was a 2-year-old patient diagnosed with lymphoblastic acute leukemia, a rare type of cancer that usually occurs in children. The prognosis is good for certain patients treated with a chemotherapy drug cocktail. From the beginning of the treatment, the patient experienced severe side effects: the number of white and red blood cells and platelets was considerably reduced. Most patients undergoing this therapy do not have such severe side effects. However, doctors were not initially aware that the treatment was causing the low counts in the blood. Scientists at the Mayo Clinic, a hospital and one of the leading centers of medical research, located in Rochester, Minnesota, discovered that carriers of a mutant gene called thiopurine S-methyltransferase (TPMT) do not produce an enzyme essential for the body to rid itself of the chemotherapy drug 6-mercaptopurine (see Table 8-3). Individuals lacking the necessary enzyme accumulate toxic levels of the drug in their bodies. The lymphoblastic leukemia patient was a carrier of the TPMT gene, and in the battle with cancer the chemotherapy cocktail was doing more harm than good. After a DNA analysis, the defective gene was identified. The doctors prescribed a more appropriate chemotherapy treatment for the patient. The dose of 6-mercaptopurine was reduced in the cocktail, allowing a more effective treatment of the disease, without the side effects of the accumulated toxins.

The next generation of pharmaceuticals being developed takes into consideration the patient's genetic makeup, so the prescription of drugs will be tailored for each individual. Therefore, the treatments will be more efficient and less aggressive for one's body.

Table 8-3
Genes and Side Effects from Pharmaceuticals

Gene	Mutant Gene Effect	Disease
CYP-2D6	Lack of capability to degrade drugs such as Prozac	Depression
TPMT	Lack of capability to degrade 6-mercapto-purine	Cancer
β-2AR	Asthmatic patients respond differently to albuterol	Asthma
ACE	Heart patients respond differently to β-blockers	Heart disease

PHARMACEUTICALS AND GENOMES . . . .

Do you realize that your medical doctor prescribes medicines just on the basis of your disease? What if he or she could decide on your prescription considering not only the illness, but your genetic makeup as well?

Genes in each human cell affect their response to drugs in two ways. Some genes code for receptors in cell membranes, which allows specific drugs to carry out their function. Other genes code for enzymes that affect the manner in which the individual is able to absorb, metabolize, and eliminate the drugs. In the former case the gene has a structural function, whereas in the latter it has a functional or metabolic role. With pharmacogenomics, the days of general-use medicines could be limited. Instead of treating breast cancer as it is done today, oncologists will determine the patient's genetic profile and establish a specially tailored therapy to maximize the efficiency of the drug in combating the disease, minimizing side effects. In other words, the medicines will become more personalized. It is likely to be many years, however, before this is commonplace in your local doctor's office.

Today, the pharmaceutical industry develops and the doctors prescribe therapies that are effective for most people, but not all. Pharmacogenomics promises to use information about the genetic differences among patients in the development of new drugs. Today, there are many deaths and complications associated with adverse reactions from medicines. Pharmacogenomics will clearly provide improvements in that area as well.

The β-2AR gene determines how well an asthmatic patient responds to albuterol, a medicine that opens the airway of the respiratory tract by relaxing the lung muscles. The gene has four or five different alleles (versions of the gene). One of them produces a high response to albuterol, whereas another reduces or prevents any response by the body. This explains why in about 25 percent of asthmatic patients, albuterol does not work properly. Knowing the patient's genetic profile, a doctor will have the ability to prescribe the right medicine, based on the disease and also on the patient. Many analysts estimate that within five years computer software will be available to help doctors prescribe drugs on the basis of their patients' genetic information. If scientific progress continues at the current pace, a genomic analysis, at an estimated cost of $500, would be the best investment one could make at a child's birth. The genomic sequence of a patient will allow a medical doctor to check genomic databases for possible side effects to medicines. Both effectiveness of the treatment and also its side effects could be foreseen prior to the beginning of a treatment. In this scenario, genes would be seen as factors that contribute to health and not only contributors to disease, as commonly portrayed.

Pharmacogenomics will eventually catalog all variations in the genome sequences across different populations. Most of those variations are single nucleotide changes from A to G or from C to T. This variation in a single base, or nucleotide, is called single nucleotide polymorphism (SNP). The current database of SNPs includes about 2 million of these variations, analyzed from individuals of several populations. Such variation is distributed in different parts of the human genome. In the studies underway, the function or association of those SNPs with an individual's susceptibility, resistance, or intolerance to medicines is being investigated. Therefore, the patient's genetic profile, composed of SNPs, might predict the response to different pharmaceuticals. With so many applications, the potential contribution of genomics for human health is still not completely fathomed by scientists in pharmacogenomics.

Although some skeptics do not agree with the potential of pharmacogenomics and the progress of medical genetics, many scientists believe that the impact of genetics in medicine will revolutionize the concept of human health. Some of the forecasts in medical genetics for the coming decades are summarized in Table 8-4.

The forecast by some scientists suggests that around 2040 the medicine practiced in many clinics will be based on the patients' own genomes. Others believe that this is much too optimistic. However, it is agreed that the scientific evolution in medical genetics is happening at a very fast pace. New discoveries occur every month. For

Table 8-4
Possible Advances in Medical Genetics in Coming Decades

Decade	Medical Progress
2010s	Genetic tests will be available in most of hospitals for 25 genetic diseases, such as colon cancer, diabetes, and many others.
	Gene therapy will be used for several medical conditions.
	Genetic medicine will be available at many health clinics.
2020s	Drugs based on pharmacogenomics will be a routine part of the treatment of common diseases such as diabetes and high blood pressure.
	Genomics will be part of the diagnosis and treatment of many diseases, including several complex disorders.
	Diagnosis and treatment of mental diseases will be based on genomics.
	Use of germ-line gene therapy.
2030s	Genes for aging will be catalogued and their function understood.
	Clinical trials for extending human life.
2040s	Integrated programs of human health based on genomics.
	Detection of susceptibility and resistance to diseases, drugs, and medicines carried out before the onset of a disease by means of genetic tests.
	Gene therapy will be available for most diseases.
	Life expectancy will be more than 90 years; many people will live more than 120 years.

example, recent studies at the NIH revealed the importance of some genes in the longevity of mice. Transgenic mice without one of these genes had a life span about 40 percent shorter than those carrying the gene. This specific gene is responsible for production of the enzyme methionine sulfoxide reductase (MrsA), which seems to be involved in mechanisms associated with protein repair. The NIH is carrying out other studies with the objective of understanding the effect of MrsA in the longevity of mice. If these results are promising, humans could be the next model for investigation.

Many people wonder why so much human health research is done in mice and pigs. No matter how incredible it seems, these animals are better models for these studies than monkeys. Their metabolic similarities to man are remarkable. Many growth hormones clinically used in humans were first developed and tested in mice. For obvious reasons, humans are not used in pharmacogenomics research until scientists have a certain degree of confidence of the safety and effectiveness of the new therapy.

Agrobiotechnology will also be contributing to the pharmaceutical industry. Beyond developing varieties that are high yielding, highly nutritious, and resistant to pests, agrobiotechnology will also be involved in pharmaceutical production. Companies such as Epicyte are focusing their efforts in agropharmacogenomics. Epicyte announced in 2001 the development of plants with antibodies for the treatment of herpes, respiratory diseases, and gastrointestinal diseases. The development of vaccines in plants, another contribution of agropharmacogenomics, is described in Chapter 13, "Bioterrorism."

HOPE FOR CURES . . . .

About 1.2 million Americans will be diagnosed with cancer in the next year. This is a fearful thing for many people. There is hope that the new drugs developed by pharmacogenomics and the new genetic tests will allow early diagnosis of this and many other diseases before they show any symptoms. Today, most children with cancer can be cured. If detected early, breast cancer can be treated without a mastectomy. Patients in chemotherapy or radiotherapy today possess an improved quality of life than those who underwent these treatments a decade ago. The hope for a cure today is substantially larger due to early detection and the existence of more efficient therapies. Clinical trials are currently in effect for various new drugs. For eligible patients these might be the best treatment available. A list of the clinical trials and other valuable information can be obtained on the Internet at *http://www.cancertrialshelp.org*.

For other information on the latest progress in human health and medical therapies, visit these Web sites:

- AllHeathNet: *http://www.allhealthnet.com*
- ClinicalTrials: *http://www.clinicaltrials.gov*
- HealthScoutNews: *http://www.healthscoutnews.com*
- HealthWeb: *http://healthweb.org*
- Cancer Treatment Centers of America: *http://www.cancercenter.com*

9 Molecular Markers

$\mathbf{A}$ common phrase in the vocabulary of many geneticists is molecular markers. Molecular markers are DNA fragments that can be used as a fingerprint in the identification or characterization of individuals. These markers have become an increasingly helpful tool in genetic research and applications to biotechnology. The basic premise behind molecular markers is that there is natural genetic variation in individuals, and many genetic sequences are polymorphic, meaning they differ among individuals. Molecular markers seek to exploit this variation to identify individuals, traits, or genes on the basis of genetic differences.

Working with the human genome, Botstein proposed the use of DNA fragments as genetic markers for monitoring segregation. The first molecular markers to be used were fragments produced by digestion of DNA with restriction enzymes. The variation in fragment size obtained from different individuals after the digestion created the class of markers called restriction fragment length polymorphism (RFLP).

One of the quickest ways to discover the location of a gene is through reverse genetics; that is, starting from the trait of interest, one would identify the protein involved. By knowing that genes code proteins, one can try to locate the actual gene. If the sequence of the amino acids of the protein is known, the genetic code can be used to establish the sequence of corresponding nucleotides, which is at least a part of the gene. From this sequence, a complementary sequence of nucleotides can be built and used as a probe. These synthetic sequences, a single DNA strand, can be used to detect genes within the billions of nucleotide bases. The probes can be radioactively labeled to facilitate identification. After the probes hybridize with the corresponding genes in the chromosomes, it is possible to identify their location by the detection of radioactivity, revealed on X-ray film. Each probe with the complementary gene can be observed on the film as a dark spot or band. Increasingly, fluorescent dyes are used instead of radioactive probes.

DNA probes are used for mapping genes in the chromosomes and for genetic tests, as in the case of the diagnosis of breast cancer. These probes are also used in the characterization of individuals at a molecular level, a process called *DNA fingerprinting*.

DNA FINGERPRINTING . . . .

DNA fingerprinting is also known as a DNA profile. This technique is useful in many areas: paternity tests, criminal cases, evolution studies, evaluation of biodiversity, mapping of genes, and genetic tests. Additionally, there are several methods for accessing the DNA profile of an individual, depending on the type of test being done.

RFLP . . . .

Developed by Alec Jeffreys in England, in the beginning of the 1980s, this technique is based on the distance between restriction sites in the DNA. The RFLP technique uses special enzymes called *restriction enzymes* to cut DNA into fragments. These enzymes recognize short, specific sequences of DNA and cut the DNA at those sites. After the DNA is treated with a restriction enzyme, it is cut into fragments of various sizes. The number and size of the fragments is unique to each individual. The restriction (cut) sites of a person, a corn variety, or a sheep are as unique as a fingerprint, allowing unequivocal identification of the individual. If the DNA of two individuals are cut with the same enzyme, *Eco*R V for example, two patterns of DNA fragments are produced, making it possible to distinguish them on the basis of the variation in the length of the fragments because each pattern of fragments is unique to each individual. The occurrence of many patterns of fragments with different lengths is called RFLP (Figure 9-1).

The relative position of bands, like a bar code, reveals the fragment sizes. The pattern of bands can then be used reliably to identify the individual source of the DNA.

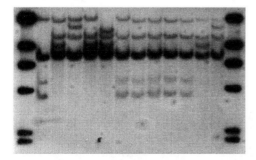

Figure 9-1
Autoradiograph of a DNA cut with the enzyme *Hind* III and revealed by a radioactive probe (RFLP).

Obtaining RFLPs

The procedure for analyzing RFLPs consists of isolating the DNA from an individual, digesting it with a restriction enzyme, and separating the DNA fragments through electrophoresis in agarose gel. The fragments are transferred to a cellulose membrane by the technique of Southern blotting. The detection of the polymorphic fragments is accomplished through hybridization with radioactive probes that can be viewed in autoradiograms.

DNA Extraction

For RFLP analysis, it is necessary to extract and purify the DNA. Purification eliminates carbohydrates, lipids, and proteins. The extraction buffer in general has a pH level of 8.0, a salt to dissociate the proteins from the DNA, a solvent to break down the lipid membranes, and a DNase-inactivating agent like EDTA (ethylenediamine tetraacetic acid).

DNA Digestion

As mentioned, DNA is digested by a restriction enzyme. These enzymes have the ability to cut DNA at precise locations in the DNA sequence. There are many possible cut sites within the genome of an organism. Most of these enzymes are derived from bacteria, which use these enzymes as part of their defense mechanism. The choice of the restriction enzymes is based on their cost and their effectiveness in producing polymorphism. Several restriction enzymes have frequently been used in RFLP analyses, including *EcoR* I, *EcoR* V, *Hind* III, *BamH* I and *Pst* I, among others.

Electrophoretic Separation of Fragments

After digestion, 2 to 15 μg of DNA are placed in an agarose gel and submitted to electrophoresis. Electrophoresis uses electrical charges to separate DNA fragments on the basis of size. After a sufficient period of time, the smaller fragments are separated from the larger ones, as the small fragments move more quickly in the electrical field. The DNA fragments are then transferred, by capillary action, to a nylon or cellulose membrane, where they are fixed in their location.

Hybridization

Nucleotide probes, short sequences of DNA conveniently labeled with radioactive P^{32}, hybridize to DNA fragments attached to the membranes, allowing them to be viewed after developing film exposed to the radiation.

The RFLP procedure requires a relatively large amount of DNA, which prohibits its use in many situations. More recently, a DNA profile generated by polymerase chain reaction (PCR) has been used to analyze samples with a much smaller DNA content, as in saliva droplets in a telephone handset or a single hair from an individual.

PCR . . . .

Kary Mullis, who won the Nobel Prize in 1993, developed the PCR technique using bacterial enzymes involved in the DNA replication for in vitro amplification of DNA. Today, PCR is being used in research as well as in laboratories, clinics, and hospitals for genetic tests, and in forensic laboratories for the analysis of DNA evidence. Many modifications have made this procedure more or less specific, as appropriate, and it is important in many genetics applications.

PCR makes copies of specific DNA sequences. A PCR reaction requires the following materials:

- DNA template: The DNA to be sampled.
- Nucleotide bases: A, T, C, and G.
- DNA polymerase: The enzyme for DNA replication.
- Primer pairs: These are short specific sequences of DNA that flank the area of DNA to be amplified. This also aids in the binding and proper function of the polymerase enzyme.
- Reaction buffer.
- Salt solution.

The procedure begins with the heating of the DNA sample in the presence of nucleotides (A, T, C, and G), a pair of primers, the polymerase enzyme, and the buffer and salt solution. Heating promotes the denaturation of the DNA molecule, separating the two strands of the DNA sample. The sample is then cooled so the primers are able to hybridize with each strand, allowing the DNA polymerase to begin the replication of DNA. As the polymerase moves along the template DNA strand, the nucleotide bases are added to create a complementary copy of the DNA template. Subsequently, the sample is heated up to again promote the denaturation of the DNA, thereby beginning a new cycle. This cycle is repeated 20 to 30 times in an apparatus called a thermocycler (Figure 9-2). The DNA area flanked by the primers is amplified exponentially.

Markers Detected by PCR

Due to its flexibility and specificity, several other methods have been and are being developed using PCR for the characterization of individual genetic variability. These include amplified fragment length polymorphism (AFLP) and SNP, as well as others. These different markers use different methods or amplify different portions of the DNA, but are used in similar PCR reaction protocols, and seek to distinguish the variation in different individuals. Some of these methods have become completely automated, such as the detection of SNP, by mass spectrometry and by hybridization on DNA microchips, which have become important tools for the development of pharmacogenomics (see Chapter 8, "Pharmacogenomics").

PCR-based molecular markers are unlike RFLP markers because they require a very small amount of DNA for analysis. The regions of interest are amplified using specialized machines and enzymes that are able to make copies of the desired DNA regions. These markers have revolutionized DNA analysis by increasing speed and reducing the labor involved for the procedures. To understand use of these molecular markers it is necessary to understand the principle of the PCR (Figure 9-2).

A series of molecular markers has been developed based on the PCR technique. Some of the markers include short tandem repeats

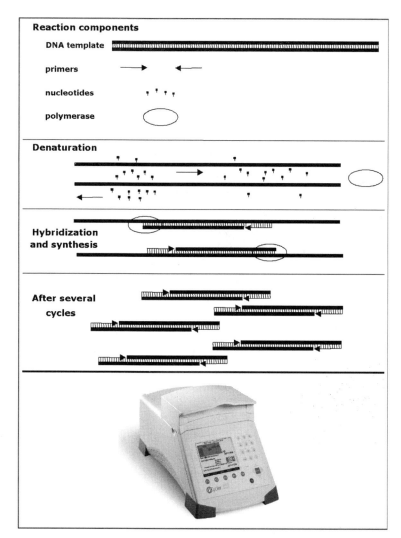

Figure 9-2
PCR reaction scheme and a thermocycler, the apparatus used for PCR.
Source: Thermocycler image courtesy of Bio-Rad Laboratories, Inc.

(STR) or microsatellites, random amplified polymorphic DNA (RAPD), SNP or point mutations, AFLP, and sequence characterized amplified regions (SCAR). They are used in studies of human, animal, plant, and microorganism genetic variability.

STR or Microsatellites

In the chromosomes of several organisms are areas containing short repeated sequence of DNA. The short sequences can be two to six nucleotides long, and they can be repeated many times. Geneticists have found them to be valuable as molecular markers and have developed PCR-based techniques to study them. They are called STR, short sequence repeats (SSR), or microsatellites. Some of the more common repetitive sequences are the dinucleotide CA and the trinucleotide CAT. These markers group in certain areas of the genome, in blocks of tandem (one behind the other) repetitions of at least five units. These repetitive blocks are highly variable and polymorphic, with the presence of several alleles due to the number of different repeated units (Figure 9-3). These areas can be easily amplified by PCR, and due to their polymorphism they are used in identification tests in criminal investigations (see Chapter 10, "Forensic DNA") and analysis of genetic variability. After PCR amplification, the products are separated by size through electrophoresis on a polyacrylamide gel (Figure 9-4). Each microsatellite can present up to two bands, in cases where there are two different alleles in the individual (one from each parent).

RAPD Markers

Two research groups independently proposed the use of small random primers in the PCR to generate polymorphic markers in 1990. This technique is known as RAPD. It uses a synthetic oligonucleotide

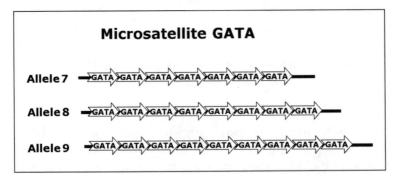

Figure 9-3
Chromosome sequence variability at the microsatellite level.

Figure 9-4
Variability in microsatellite markers for a
wheat disease-resistance gene.
Source: Courtesy of Dr. Sixin Liu, University
of Minnesota.

as a primer in the amplification process to produce a polymorphism
detected by the presence or absence of a band. This technique consists of extracting the DNA of the individuals to be analyzed and
using this DNA in the amplification reactions (PCR), which can be
done using a different primer for each reaction. The products of each
reaction are a result of the amplification of different chromosomal
areas flanked by the pair of primers. The amplified fragments are separated by electrophoresis in a gel. Different individuals produce different patterns of amplified fragments, or DNA profiles (Figure 9-5).

Among the markers currently in use (see Figure 9-6), the RAPD
technique is simpler and quicker than RFLP. It uses smaller amounts
of DNA, does not involve the use of radioactive probes, and is less
labor intensive. It has been used mainly in plants because it does not
detect variation in humans and other animals.

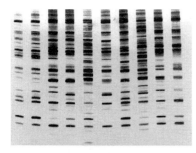

Figure 9-5
Gel with RAPD patterns from different
individuals.

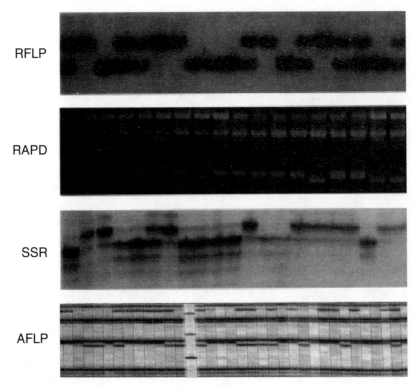

Figure 9-6
Different molecular markers used today.

10 Forensic DNA

In this chapter...

In 1892, fingerprints began to be used in personal identification. The discovery that an individual's fingerprint is a unique characteristic (i.e., no other person would have the same pattern in the population) stimulated their use for identification purposes. Fingerprints are so unique that not even identical twins possess the same fingerprints.

The judicial system has been able to convict a great number of suspects from fingerprints left on surfaces such as furniture, telephones, and glass. On the other hand, a comparable number of individuals have also been cleared of criminal charges because of this identification tool.

Similarly, molecular individual identification, or DNA fingerprinting (see Chapter 9, "Molecular Markers"), has become widely used to solve forensic cases. The first significant case of the use of DNA as criminal evidence occurred in 1986 in England, when a homicide suspect was released after DNA analysis of evidence collected at the crime scene was compared with the suspect's DNA fingerprint. Beginning in 1987, the U.S. Federal Bureau of Investigation (FBI) and several other criminal laboratories in different countries began to use DNA as biological evidence in criminal cases. Samples of blood, saliva, hair, semen, and other human cells found at the crime scene became strong evidence for the imprisonment or release of suspects.

As seen in Chapter 9, molecular markers can be used to characterize an individual's DNA in a pattern or profile of fragments that is unique to him or her. As opposed to normal fingerprints that can be altered by surgery, a person's DNA cannot be deliberately altered. Consequently, the DNA profile has been considered an important and reliable method of individual identification.

The genetic information contained in DNA is determined by the sequence of the letters of the genetic alphabet (A, C, G, and T). In humans, about 3 billion of these letters are arranged in the same order in the chromosomes of each cell in the human body. It is the order in which the letters are arranged in the chromosomes that makes each individual unique from all others. Obviously, the more dissimilar the individuals are, the more distinct is the order of the nucleotides (letters) in the genome. Similarly, individuals who are genetically related (e.g., siblings, parents, and children) have proportionally larger similarity in their gene sequences. Ultimately, only identical twins have the same DNA sequence. Therefore, a DNA profile is a simple and fast way to compare DNA sequences of two or more individuals.

DNA profiles have many varied applications. They can be employed in criminal and civil cases, used for paternity tests, help in determination of succession of properties by inheritance, used for identification of bodies, or used to determine property rights of crop varieties, among other things.

In December 1954, Sam Sheppard's trial in Cleveland, Ohio, occupied the media across the United States. Convicted to life in prison for the murder of his pregnant wife, Sheppard repeatedly maintained his innocence. In 1966, the Supreme Court threw out the trial because of trial errors. Sheppard was freed, but he died four years later. In 1992, a book was published accusing Richard Eberling, a neighbor of Sheppard, of committing the crime. Eberling had been convicted of murdering another person and he died in jail. After his death, other inmates advised authorities that Eberling had admitted to the murder of Sheppard's wife. In 1997, Sheppard's son requested authorization to exhume his father's body with the objective of obtaining a DNA analysis to substantiate his father's claims of innocence. The analysis indicated that his father's DNA did not correspond to the evidence collected at the crime scene, refuting the possibility that Sheppard had committed the crime.

Between 1989, when the FBI began to use DNA analysis in rape cases, and 1996, the agency investigated about 10,000 cases of sexual abuse. DNA tests excluded about 25 percent of the primary suspects in those cases. In many criminal cases, eyewitnesses, or more frequently the victims themselves, are needed to positively identify the suspect. DNA has shown those eyewitnesses are not always reliable. When properly collected, manipulated, stored, and analyzed, DNA has the potential to eliminate numerous errors in the criminal justice system. DNA is valuable in all situations, and even more important when eyewitnesses cannot be found. It is almost impossible to commit a crime without leaving DNA evidence such as hair, skin cells, and body fluids at the crime scene.

The Innocence Project at the Benjamin Cardozo Law School in New York is a clinical law program for students supervised by law professors and administrators. The Innocence Project provides *pro bono* legal assistance to inmates challenging their convictions based on DNA testing of evidence, although clients must obtain funding for the testing. Founded in 1992 by Barry Scheck, Professor of Law, and Peter Neufeld, the Project has represented or assisted in many cases in which convictions have been reversed or overturned in the

United States. To date, this project has been responsible for the exoneration of more than 100 people.

DNA analysis is transforming biological evidence as an irrefutable instrument for incrimination or absolution of suspects. It eliminates some of the ambiguities of the justice system. Even so, in the controversial case in which O. J. Simpson was charged with murdering his ex-wife Nicole Brown, the evidence failed to provide conclusive results. In a highly disputed outcome, Simpson was acquitted of the charges, in part because his lawyers challenged all biological evidence. As in the Simpson case, skilled lawyers are always challenging the chain of custody, hoping to convince juries that DNA evidence has been contaminated. As with all evidence collected, an important requirement for the validity of DNA tests in criminal trials is the integrity of those who collect, process, and store the evidence.

DNA ANALYSIS . . . .

The DNA digestion with restriction enzymes produces fragments of different lengths, according to the individual's genome sequence. For instance, an individual that has the sequence AAGCTT, the cut site for *Hin*d III, will have his or her DNA cut by this enzyme in as many fragments as that sequence occurs. Therefore, if the DNA of a suspect S_1 has 50 restriction or cutting sites, whereas suspect S_2 has 55 restriction sites, the fragmentation of the DNA of the two individuals will produce different patterns. The size of each fragment depends on the distance between sites, and the number of fragments depends on the number of cut sites. This variation of fragment number and size is usually referred to as polymorphism, due to the multiple forms in which the DNA is cleaved.

Obtaining a DNA Profile

The DNA profile of an animal, plant, or microorganism can be obtained by analyzing its DNA.

A simplified protocol might include seven steps:

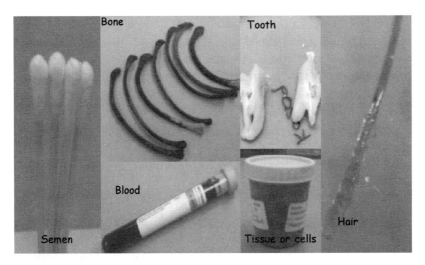

Figure 10-1
DNA sources used as criminal evidence.
Source: Courtesy of the Minnesota Bureau of Criminal Apprehension.

1. Harvesting the biological sample: Blood, saliva, semen, hair, tooth, bones, or any other cellular tissue or fluid from the individual (Figure 10-1).

2. DNA extraction: DNA should be isolated from the sample. Depending on the preferred method, a tiny amount of sample might be sufficient, such as the defoliated cells from the skin of an individual's forehead found in a hat. Alternatively, a saliva droplet left on a telephone handset or postage stamp might also contain enough DNA for the analyses.

3. DNA digestion: The following step is the DNA cleavage with a restriction enzyme. The enzymes *Hin*d III and *Eco*R I, among others, have frequently been used for this purpose. After the DNA is treated with the restriction enzyme, it is then made up of a number of fragments of different sizes.

4. Fragment separation: The DNA fragments are separated by size using electrophoresis. This procedure consists of submitting the DNA fragments, inside of a gel, to an electric field. The gel is usually made of agarose, a substance

extracted from sea algae that is similar to gelatin in composition. The DNA is placed in a small well on the gel close to the negative electrode. An electric field causes the DNA fragments to migrate toward the positive electrode. Small DNA fragments move quicker than large fragments, allowing separation by size.

5. DNA transfer: After the separation of the fragments, they are transferred to a nylon membrane by capillarity. Once they are attached to the membrane, the fragments can be manipulated for viewing.

6. Hybridization with probes: The addition of fluorescent labeled or radioactive probes to the nylon membrane allows the visualization of the fragments that are complementary to the probes. Each probe typically highlights some of the fragments in the membrane.

7. DNA profile: The final DNA profile is obtained after the hybridization of various probes to the membrane. The result is a pattern of bands (dark spots) of different sizes (Figure 10-2).

An Example

DNA as criminal evidence or as a means for individual identification is revolutionizing law enforcement. Consider the following example, in which two individuals, S_1 and S_2, are rape suspects. Analysis of the semen or any other tissue collected at the crime scene, here designated E, can be used to identify the individual responsible for the crime without submitting the victim to the additional stress of testifying at the trial. This is especially important when the victim is unable to identify his or her aggressor.

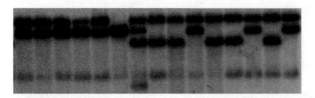

Figure 10-2
DNA profile showing fragments of different lengths.

The DNA analysis of the two suspects and of the biological evidence can provide unmistakable evidence for the true perpetrator of the crime. After DNA analysis, one of the possible profiles is presented in Figure 10-3. Examining the profiles produced with the four probes in this figure, suspect S_1 can be excluded as the instigator of the crime because his DNA profile is different from that of E (probes 1, 2, and 4). Technically, it would be more accurate to say that suspect S_1 could be cleared of having left the biological evidence at the crime scene. Good criminologists understand that the relationship between the evidence, the crime, and its authorship should receive equal attention. However, it is clear from the DNA analysis that E does not have the same profile as suspect S_1.

The DNA of suspect S_2 corresponds perfectly with that of E in the four regions of the DNA analyzed with the four probes. That doesn't necessarily mean that the suspect S_2 is the author of the crime. Obviously, to know with what certainty the evidence should be considered, it is necessary to know with what frequency that same DNA profile is found in the human population, and that is a matter of probabilities. After the DNA analysis with a certain number of probes, the laboratory produces a report indicating the probability that the DNA of suspect S_2 and E would be the same by coincidence. Usually, the level of statistical stringency is extremely high to prevent costly errors.

Besides RFLPs and RAPDs, other types of molecular markers have been used in DNA analyses for criminal purposes. For instance,

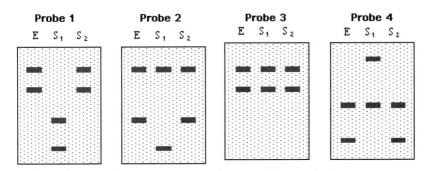

Figure 10-3
DNA profile of two suspects, S_1 and S_2, and of the evidence E collected at a crime scene.

several forensic laboratories, including the FBI, are using STR in their analyses. Those markers are highly discriminatory, and the analysis of 13 regions of the human genome, with markers now available, allows the generation of reports stating the probability of up to 1 in 82 billion individuals.

Databases

DNA profiling is a powerful identification tool. Most countries have fingerprint databases that are used in criminal cases. In crimes against properties, fingerprints can often be found. However, in violent crimes, fingerprints are not typically left at the crime scene. Rapists usually touch the victim's body, leaving no fingerprints. However, semen on the victim's clothes or body can be used as evidence for solving such cases.

Currently, many countries are establishing DNA databases from individuals with a history of violent crimes. Those databases would allow the identification of suspects by simply cross-checking the DNA profile of the evidence with those stored in the database. However, several ethical issues have been raised about those databases. More details on DNA and genetic privacy are discussed in Chapter 14, "Bioethics."

Reliability of DNA Tests

Although a DNA profile is considered irrefutable proof of identification, it is necessary to establish standards of analysis and accuracy levels in the statistical calculations. Additionally, the laboratories that provide these services should be submitted to double-blind tests, in which neither the lab nor the technicians know what the samples are, to ensure that they are working with acceptable quality control.

Analysis of Mitochondrial DNA

A different approach has been given to the analysis of DNA in mitochondria, an organelle inside the cells. In cases where the amount of DNA is extremely small or in cases where the victim has been carbonized, partially destroying the nuclear DNA, the analysis of the mitochondrial DNA is an alternative. However, it should be acknowledged that mitochondrial DNA tends to be much less variable, because the mitochondria is maternally inherited; that is, an individual's mitochondrial DNA is exactly identical to that of his or her mother and siblings, and to all uncles from his or her mother's side of the family, and so forth. The use of mitochondrial DNA for individual identification was extremely important in Argentina in cases in which children were separated from their families during a military dictatorship. The movement known as "The Grandparents of May" in Argentina was able to reunite many children with their blood families on the basis of mitochondrial DNA analysis. The same analysis was also used in the identification of some bodies from the World Trade Center terrorist attack on September 11, 2001 in New York City.

PATERNITY TESTS . . . .

DNA tests are the most accurate and reliable technology used for paternity identification today. Usually, samples from mother, child, and alleged father are tested to determine if the alleged father is the biological father of the child.

The paternity test report should clearly indicate one of these two alternatives:

1. The tested individual is excluded and, therefore, he or she cannot be a biological parent of the child.
2. The tested individual is not excluded as the biological parent of the child. The statistics in this case should indicate the probability that the alleged individual can be a biological parent.

Accuracy of the Tests

Paternity tests can prove with complete certainty that an individual is not a child's biological parent. However, there is no available test that can prove with 100 percent certainty that an individual is the child's biological parent. Paternity tests can guarantee at most 99 percent probability of paternity.

The following situations can result in reduced precision of the tests:

- Interracial marriage or nonstandardized populations. This basically means it is more difficult to use highly specific DNA fingerprinting methods because of the large sample population. In those cases, the laboratories are forced to use generic tables in the analysis.
- A biological parent is related to the alleged parent. Such cases are more serious in close relationships as in brothers, or father and son. The paternity test cannot exclude an individual when the alleged parents are identical twins.

A paternity test report should provide the following information:

- Identification of the tested individuals and their pictures
- Details of the procedures used for collecting of the biological samples
- Description of the procedures of confidentiality and processing of samples
- Description of the molecular techniques used in the test
- Identification of the locus used in the test and of the alleles found in each tested individual
- Number of analyzed loci
- Reference to the genetic database used in the analysis
- The digitized pictures of the genetic analyses that prove the presented results
- In cases of exclusion, the paternal alleles not expected to be found in the alleged parent that indicate that he or she could not be the child's biological parent
- In cases of inclusion, the three statistics of paternity used

The same considerations for individual identification presented in this chapter apply for the identification of both animals and plants. The same methods used in forensic DNA testing are also used in patent litigation to determine property rights for some crop varieties. The DNA profile of hybrid corn varieties was used in a long and controversial lawsuit between two seed companies in the United States that argued over the improper acquisition of parental inbred lines. The DNA analysis of the inbred lines was used to settle the case. Today, many seed companies use molecular identification for its elite germplasm.

GENOMIC PICTURES . . . .

Considering that all phenotypic characteristics, such as color of eyes, hair, skin, face format, and so on, are defined by the genome, many forensic scientists believe that in the future it will be possible to substitute for the composite picture drawn based on information reported by an eyewitness with a composite drawn on the basis of the analysis of biological samples left as evidence at the scene of a crime (see Figure 10-4). Some still speculate that in the near future computer software will compose the individual's picture automatically, considering the individual propensity for obesity, baldness, and other characteristics. However, an individual's physical appearance is the result not only of his or her genes, but also of environmental factors. Many characteristics are affected for a great number of genes. This type of technology is a long way off, but the unveiling of the human genome is allowing scientists to gain a better picture of how DNA influences each one of us, and is facilitating the recognition of each unique individual.

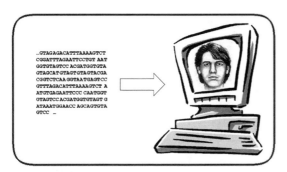

Figure 10-4
Genomic picture: Reality or dream of forensic science?

FINAL CONSIDERATIONS . . . .

Considering that DNA analysis is a powerful identification technique, it should be used carefully. The sensitivity level of many DNA tests is so high that cells from a technician's hands or from a sneeze could contaminate the sample.

Therefore, care in the collection, custody, and manipulation of the biological sample is of great importance for the validity of these analyses. Finally, human beings can make mistakes. Technicians can mislabel a flask, change codes, change names, and so on. Due to the many possible errors, many laboratories use double reading in each step of the analysis. They also save part of the sample for possible re-analysis. Even so, mistakes will continue to happen, and in many cases it is left to skilled lawyers to question and criticize unexpected results.

Some scientists believe that in the future the DNA profile will be part of the personal identification used in identity cards (Figure 10-5).

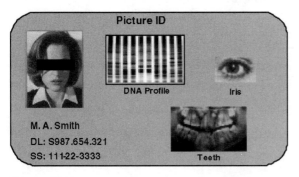

Figure 10-5
Identity card of the future?

11 Bioremediation

In this chapter...

With an increasing emphasis on the quality of our environment, many recognize that science can play a vital role in the improvement of the quality of our air, soil, and water. Bioremediation is the use of biological systems to degrade or remove noxious pollutants from the environment. Biotechnology is becoming important in this field as well.

Environmental pollution is the result of population growth and technological progress. Short-sightedness has blinded humans to their effect on the environment. In the race for progress, society values productivity while ignoring waste and garbage left along the road.

The main objective of technology is to improve the well-being of mankind. The benefits of technology to man are easily seen, one example being the increase in life expectancy from 40 years at the beginning of the 20th century to more than 70 years today. However, technology has also inflicted a high price on the environment. Pollution can be seen or felt everywhere. It is not absurd to think that the technology advancements that have contributed to the problem might also be the source of the solution.

DEPOLLUTING ORGANISMS . . . .

As Lavoisier said when enunciating the Law of Conservation of Mass, "In nature nothing is created and nothing is lost, everything is transformed." In reading this chapter it becomes evident that nature offers alternatives for transformation of pollutant residues into non-pollutant ones by the action of microorganisms and plants.

Microorganisms

In 1975, a leak in the jet fuel tank of a U.S. Air Force airplane contaminated the soil in Charleston, North Carolina. More than 300,000 liters of the chemical toluene leaked into the soil, reaching the underground water supply. By 1985, the contamination had spread to many residential areas. The removal of the polluted soil was technically impossible, and the removal of just the contaminated water would not solve the problem; it was impossible to eliminate the source of pollution, the soil. One possible solution was to use natu-

Figure 11-1
Different bacteria have the capability to transform different substances into energy for their growth.

rally occurring microorganisms in the soil that have the capability to digest toluene, transforming it into carbonic gas, water, and energy (Figure 11-1).

In 1992, bioremediation was used in Charleston, North Carolina, to clean the environment. Scientists added nutrients to stimulate growth and activity of specific microorganisms that would degrade the fuel. The nutrients were applied in the contaminated soil through infiltration tubes, and the polluted water was removed from several artesian wells. About a year later, the level of contamination was reduced by 75 percent. Close to the infiltration tubes, where the population of the bacteria had higher growth rates, the previously high levels of toluene had been reduced to undetectable levels.

Bacteria were also used for cleaning the Alaskan coast in 1989, after the infamous Exxon *Valdez* oil tanker ran aground, spilling tons of crude oil (Figure 11-2). Most of the viscous oil was removed from

Figure 11-2
Oil pollution has an enormous impact on the environment.

Source: Tanker photo courtesy of Damage Assessment and Restoration Program, National Oceanic and Atmospheric Administration.

the sea by suction and filtration of the superficial layer of oil and water. However, the oil that penetrated rocks and gravel along the beaches was cleaned by bacteria that had the ability to decompose crude oil.

The bacteria used for cleaning the Exxon *Valdez* spill used oil, a hydrocarbon, as source of energy for their growth, decomposing it into smaller nontoxic compounds. Various bacteria possess different abilities to decompose residues that are toxic to man. The diversity of microbes is enormous. Whereas some microorganisms survive by feeding off of other living cells, many others use decomposing organic matter for their survival. There are also many that are able to use toxic substances as a source of energy. Some microorganisms have very eclectic metabolic pathways that allow them to use solvents such as chlorine, a typical pollutant in highly industrialized areas, as a source of energy. Those microorganisms can use chlorine as an oxidant when oxygen is not available.

Several research groups have identified bacteria that can degrade agro-chemicals and chemical fertilizers. This is additional evidence of the importance of the preservation of biodiversity. A seemingly useless microorganism in an environment might eventually be used for cleaning that environment. With the development of biotechnology, microorganisms are drawing attention from scientists and biotechnology companies. The invisible army of microorganisms that continually promotes recycling in the Earth is acquiring a new prestige.

Experience with bacteria seems to indicate that the microorganisms can use practically any substance as "food" or a source of energy. Although some materials are highly toxic for a certain bacterium, they can be a substrate for another. Bacteria have been isolated that are able to feed on detergents, sulfur, methane, chlorine, carbon tetrachloride, toluene, and other substances. The great biodiversity of microorganisms on Earth is a largely untapped reservoir of species with unexpected abilities.

Methylene chloride is considered one of the most serious pollutants because of its carcinogenic properties. This substance is produced in large amounts in certain industrial processes. It is, however, decomposed into water, carbonic gas, and salt when treated in bioreactors with a species of bacteria that decomposes the methylene chloride using enzymes to convert the chemical into energy and other nonpollutant substances. The discovery of microorganisms for cleaning the environment, in general, occurs in places where there is pollution. Microorganisms found growing in those places are at the very least resistant to pollutants occurring there. Bioremediation typically consists of harvesting many microorganisms from polluted sites to select those with the most efficient degradation abilities. They are then isolated, multiplied, and reintroduced into the areas to be cleaned. In addition to the microorganisms, some nutrients, such as nitrates and phosphates, are added to the inoculum to promote fast growth and enhanced cleaning. Depending on the nature and extension of the pollution, the bioremediation can take from a few months to many years to work. When the toxic substance is eliminated, the population of the microorganism is also reduced. Eventually, other populations of bacteria will find favorable conditions of substrate, temperature, and humidity, and will be able to grow.

One of the largest hurdles with bioremediation is the difficulty in controlling the factors for the microorganism growth. The development of a microorganism is affected by temperature, pH, humidity, and availability of an energy source. It is relatively simple to culture bacteria in a laboratory under controlled conditions, but outside in changing environmental conditions, the process is more difficult. To assist in this effort, biotechnology is establishing a constructive partnership with bioremediation to engineer more efficient microorganisms that are less dependent on environmental conditions. For instance, transferring genes from thermophilic (heat-loving) bacteria to one that can decompose insecticides would allow the transgenic microorganisms to be used in areas in more extreme temperatures. Several research groups are developing genetically modified bacteria that have enhanced capacity for cleaning areas polluted with heavy metals, radioactive elements, chemical fertilizers, insecticides, herbicides, and other toxic elements (Figure 11-3).

Degradation of Radioactive Compounds

Uranium (U) is the most common contaminant at facilities of the U.S. Department of Energy. The uranyl ion $[UO_2]^{2+}$ is a common, soluble form of this element in the environment. Microbes can immobilize the uranyl ion in several ways, three of which are shown in Figure 11-4. The mineral uraninite (UO_2) is highly insoluble. Microbes can reduce uranyl ion into hydrated uraninite. A cytochrome-c3 hydrogenase

Figure 11-3
Bacteria from soil contaminated with herbicides.
Source: Courtesy of Dr. Julieta Ueta, Pharmaceutical
Sciences Department of the College of Pharmaceutical
Sciences of Ribeirão Preto, Brazil.

from the *Desulfovibrio vulgaris* bacteria and other organisms can
carry out the reduction. Reaction A in the middle branch of Figure
11-4 can be carried out by *Deinococcus radiodurans*.

The uranyl ion can also be precipitated as cell-bound hydrogen
uranyl phosphate without a change in the oxidation state of the uranium, as shown in the right pathway branch in Figure 11-4. This reaction is facilitated by acid phosphatase N from *Citrobacter* sp.

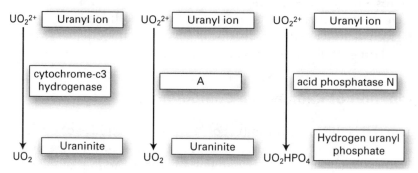

Figure 11-4
Uranium biodegradation.

Depolluting Plants

Aside from the world of microbes, some plants have also been found to possess the ability to absorb and compartmentalize different elements from the soil, making them useful in bioremediation. At the beginning of the 20th century, it was speculated that some plant species could be used as indicators for the presence of gold in the soil. Populations of certain species were especially abundant in areas rich in the metal.

The mobilization and concentration of metals in plants involves a group of proteins called *metalthionins*. Some of those proteins are very selective, accumulating specific types of metals. The capability of some plants to absorb metals can be used to extract heavy metals from polluted soil and water.

For a plant to be used in bioremediation, it should do the following:

- Mobilize the metal, or make the metal available for uptake.
- Absorb the metal with its roots.
- Transport the metal from the roots to the canopy.
- Retain the metal in its tissue.

Plants should absorb and accumulate the metal in an efficient way and must be adapted to a wide range of environmental conditions so that they can be introduced in the most diverse environments where pollution might occur.

After cleaning the polluted area, plants will have absorbed the contaminating element and completed their life cycle. They should then be harvested and removed from the area, and the pollutant present in their tissue should be treated in a way to avoid pollution elsewhere. Biotechnology offers opportunities for improving the capability of mobilization and absorption of metals by plant species by the overexpression of the genes that code for metalthionins. Transferring this trait from bacteria or a plant to a crop species with large adaptability could be one of the major contributions of genetic engineering in bioremediation.

FINAL CONSIDERATIONS . . . .

A clean and healthy environment has strong appeal in all aspects of life. A great part of the soil, water, and air is already polluted, and steps are being taken to clean up these problems. The current challenges for bioremediation are to reduce or to eliminate pollution added to the ecosystem and to clean the contaminated areas. Biotechnology might assist in addressing both challenges. As previously seen, the development of transgenic organisms that have the ability to degrade pollutant residues before they are released in nature will probably be a remarkable contribution to modern society.

At present, the practical use of transgenic microorganisms for bioremediation has been limited to a few field tests. Their future use is still subject to further research to understand their behavior in different environments. Therefore, it is still necessary to acquire basic knowledge about the behavior of genetically modified cleaning organisms in complex environments. Although the promises of biotechnology for remediation of pollution are auspicious, caution is still needed, as some of the promises could take years to provide results. It is important to recognize that bioremediation is not an excuse for society to continue irresponsibly damaging the environment. Besides, it is widely accepted that prevention of pollution is less expensive and more politically sound than the adoption of cleaning strategies.

12 Biodiversity

In this chapter...

Biodiversity, or biological diversity, refers to every form of life within an area or ecosystem. This includes the genetic variability within the populations and species; the different species of flora, fauna, and microorganisms; the variety of functions and ecological interactions carried out by the organisms in the ecosystems; and the various communities, habitats, and ecosystems formed by the organisms. Biodiversity is the fruit of the great laboratory, which is the planet Earth, with its more than 30 million different species resulting from 4 to 5 billion years of evolution.

The importance of preserving biodiversity is also referred to in sacred books, such as the Bible, which relates that Noah saved domestic and wild animals from the great flood. Biodiversity is one of the fundamental properties of nature responsible for the balance and stability of ecosystems. It is also of great economic value. This diversity is the basis of farming and food production, and it is essential for biotechnology. The ecological functions carried out by various organisms are still poorly understood, but biodiversity is thought to be responsible for the natural processes and products supplied by ecosystems. It accounts for the species that sustain other life forms and also modifies the biosphere, making it suitable and safe for life. Biological diversity possesses, besides an intrinsic worth, a value of ecological, genetic, social, economical, scientific, educational, cultural, recreational, and aesthetic importance.

A reduction in biological diversity is hazardous to sustainable development. Genetic erosion (the loss of species variability) and the extinction of species can influence us to develop strategies that contribute to the preservation of the remaining biodiversity on the planet, at a level that is already smaller than it was a century ago. The preservation of biodiversity is also essential for human wellbeing. However, recent studies have indicated that extinction rates are 1,000 times faster than those expected naturally, with 50,000 species extinguished every year. Currently, about 34,000 plant species and 5,200 animal species are at risk of becoming extinct.

Biotechnology can be understood as a technology that explores biological systems instead of individual living organisms. Therefore, the preservation of the biological systems with all of their diversity can be considered a priority as well as a challenge to mankind.

Microbes, such as bacteria, are the most diverse of all living organisms. Some estimates indicate that there exist more than 1 million different species of bacteria in the world. Recent reports suggest that an extremely large number of bacteria exist in the biosphere awaiting the development of appropriate techniques needed to grow them, so that they can be characterized.

This is one example of one key part of the greater picture of biodiversity. Plants, animals, and even fungi are also important aspects of the world's biodiversity. This idea of biodiversity is an important part of biotechnology, as useful traits and chemicals are becoming part of important new biotechnology applications. Biotechnology brings, simultaneously, promises of biodiversity preservation and also the fear of genetic erosion and biopiracy.

PRESERVATION OF BIODIVERSITY . . . .

The preservation of genetic variation has become an important subject for many species. Various species of plants, animals, and microorganisms have been collected and stored, so the immense species variation might not be lost. The culture of cells and tissue, an area within biotechnology, is being used for the maintenance of live collections of the most varied types of plant species of economic importance or others at risk for extinction. For instance, the preservation of the genetic diversity of cassava is accomplished at tissue culture laboratories, where thousands of different varieties and species are maintained in small petri dishes. In the Frozen Zoological Garden in San Diego, California, there are live cellular lineages of species of several families of mammals, many close to extinction. It is expected that in the near future, cloning techniques will be used to regenerate whole animals from the cells. Had tissue culture technologies not been developed, the required space and costs to preserve rare species would be many times larger, limiting the number of species that could be preserved.

The preservation of microbial diversity has also been made possible by biotechnological techniques. If the bacteria, fungi, and viruses had to be maintained in their traditional hosts, only a small fraction of the biodiversity of microorganisms could be preserved. The germplasm banks of bacteria and fungi require a relatively small and rudimentary laboratory for preservation. The main objectives of microorganism gene banks are related to the preservation of species for subsequent laboratory studies.

GENETIC EROSION . . . .

Until the 1940s, the centers of origin of crop species and animals were considered limitless sources of genetic variability. After World War II, agriculture in developing countries suffered great changes. The expanded use of improved varieties resulted in the reduction of traditional varieties, a process called *genetic erosion*. The expansion of the agricultural frontiers also contributed to the risk of loss of the wild relatives of crop species.

According to a study carried out by the National Academy of Sciences in the United States, of approximately 3,000 possible plant species, only 20 to 30 constitute the basis of agriculture. For example, amaranth has high economic potential and has been recommended as a species that deserves more attention from plant breeders, with the objective of improving the plant to make it more valuable for commercial use. This requires the removal of undesirable traits and the improvement of other traits to allow for improved production.

The process of genetic erosion also occurs with many other species of flora, fauna, and microorganisms, and it is the first sign indicating possible species extinction. Environmental deterioration initially results in local extinction and later culminates with the global extinction of the species. For instance, well before species vanish, a small number of survivors could result in inbreeding of the population. Inbreeding results from intermating between related individuals that causes the generation of less fit individuals with a greater likelihood of genetic defects. Biotechnology can help in the diagnosis of genetic erosion before any conventional techniques. This can be achieved by DNA analyses that quantify the remaining genetic diversity (see Chapter 9, "Molecular Markers"). However, as each organism has a different genome, these methods would have to be developed for each species. This technique has been used with success in the study of wolf species, fish, cattle, macaws, whales, and other animals. In many cases, the studies were used to justify the creation of new refuges where such species dwell.

GENE BANKS . . . .

The term *germplasm* is vague and imprecise. However, germplasm has been defined as the entire hereditable material or the whole genetic makeup of an entire species. In other words, germplasm is all the existent diversity for a considered species. It is essentially biodiversity at a genetic level.

Two basic methods exist for germplasm conservation: ex situ and in situ. Gene banks function as ex situ conservation, where a sample of the genetic variability of a species is preserved in an artificial environment, outside of its normal habitat. In general, the seeds of plant species are stored in environments at low temperature and humidity. In these conditions, their viability can be preserved for several decades. Several gene banks around the world function as centralized storage areas for the germplasm of certain species. For instance, the USDA has a facility in Fort Collins, Colorado, that was specifically designed for long-term storage of important plant species. Other locations house working collections of germplasm from more specific species, such as the National Small Grains Germplasm Research Facility in Aberdeen, Idaho, which holds a working collection of wheat, barley, oats, rye, rice, triticale, and wheat relatives. This and other storage areas maintain germplasm diversity while allowing researchers to use the materials for plant improvement purposes. When there is a need for more long-term preservation, such as for preserving some specific types of plant tissue used in tissue culture or preservation of pollen, cryopreservation in liquid nitrogen at −196°C is the appropriate alternative. This is the storage method used in the Fort Collins National Seed Storage Laboratory.

In the in situ collections, the germplasm is preserved in its natural habitat. Many in situ gene banks are not recognized as such, mainly due to the terminology they receive (i.e., biological reserve, national park). Obviously, not every national park is a gene bank if the preservation of resources is not monitored, or if it is created for preservation of an entire ecosystem and not a certain species. A typical example of an in situ gene bank is located in Mexico, where 140,000 hectares in the mountains of Manantlan were designated as a biological reserve of *Zea diploperennis* (perennial diploid teosinte), a wild relative of maize. Its population is monitored periodically to detect any risk for loss of genetic diversity.

The importance of biodiversity becomes most evident during crises, as when a plant disease epidemic occurs, varieties of a crop species are susceptible to the pathogen, and no other resistant varieties exist. In these situations, the plant breeder usually looks to gene banks or at the centers of diversity of the crop for resistance sources that can be used in their crop breeding programs. Several varieties of rice, tomato, potato, and other species were developed using genetic resources from gene banks.

BIOPIRACY . . . .

Plants constitute a rich source of therapeutically important products. Only 10 percent of plant species have already been tested for pharmaceutical value. About 120 medicines commonly prescribed by doctors are based on plant extracts. Some of the most important medicines used by humans have a history that traces back to medicinal plants collected from wild flora, and others are just now beginning to be discovered.

Aspirin is one example of the health benefits provided by plants. This medicine is based on acetyl-salicylic acid, a very common analgesic. Its history began with the Greek doctor Hypocrites, who in the fifth century B.C. used a bitter powder to treat pain and lower fever. This mystical powder was collected from the cork of *Salix* (Figure 12-1), a tree of the family *Salicaceae*. Although the mode of action of aspirin was only unraveled in the 1970s, this medicine has been used as a painkiller and to improve the elasticity of the circulatory system for millions of people. If this species had gone extinct before that discovery, man would have never known the valuable medicine. In fact, Americans annually consume about 80 billion aspirin tablets.

Another example of the use of a plant extract is eye drops used for treatment of glaucoma. The active ingredient of the eye drops, pilo-

Figure 12-1
Salix sp., species from which aspirin was originally
extracted.

carpine, is extracted from *Pilocarpus pinnafolius*, a native species of
northeast Brazil. Pilocarpine was originally used by native Brazilians
to induce sweating (Figure 12-2). A pharmaceutical company owns a
farm of 3,000 hectares in Barra da Corda, Brazil, of which 400
hectares are planted with 15 million *Pilocarpus pinnafolius* trees for the
production of pilocarpine salts.

N-Tense is another example of a therapeutic product made from
medicinal plants in the Amazon rainforest. The plant extracts in this
formula have been documented around the world to have antitumor-
ous, antibacterial, anticancerous, immunostimulant, and antiviral
properties (Figure 12-3). Some of the species used in this product
include *Physalis angulata, Annona muricata, Scoparia dulcis, Maytenus
ilicifolia, Guazuma ulmifolia,* and *Momordica charantia,* among others.
These are just a few of several herbal medicines reported in the
literature.

The history of biodiversity collection dates back to 1500 B.C., when
Egyptian rulers gathered plant species from their military expeditions. Charles Darwin, the renowned naturalist of the 19th century,
accomplished one of the most famous trips for biological collection.
During his travels on the ship the *HMS Beagle,* he collected samples

Figure 12-2
Pilocarpus pinnafolius produces pilocarpine,
an active ingredient in the treatment of
glaucoma.
Source: Image courtesy of Raintree Nutrition Inc.,
www.raintree.com.

of everything that interested him, from which he elaborated the *The-ory of the Evolution*, the foundation of modern biological research. More recently, Nicolai I. Vavilov, a Russian scientist in the beginning of the 20th century, also collected samples of plant species from five continents, with which he established the *Theory of the Center of Ori-gin of the Crop Species.*

None of those famous expeditions were legally or morally ques-tioned. Today, the paradigms and laws have changed, and biopiracy is considered a crime. *Biopiracy* is the unauthorized appropriation of

Figure 12-3
N-Tense, produced with plant extracts
from Amazonian species.
Source: Image courtesy of Raintree Nutrition Inc.,
www.raintree.com.

any biological resources. The extraction of aromatic, ornamental, or medicinal plants without the proper authorization is considered biopiracy. Natural resources primarily from Africa and South America are becoming increasingly valued in the international market. In the 1500s, Brazilian wood was prized for making red dyes; today, targeted Brazilian species number about 50,000 plant species, 534 mammals, 3,000 fishes, approximately 1,700 birds, 500 amphibians, and 470 reptiles. The wealth of Brazilian biodiversity makes the country a valuable source of genetic diversity. Every year, thousands of tourists, scientists, environmentalists, and biologists travel around the world under the umbrella of ecological tourism. Although this type of travel has improved research on biodiversity, it has also caused problems relating to biopiracy.

Scientists and pharmaceutical companies are obtaining several patents using plant extracts from different regions around the world. Recently, the English chemist Conrad Gorinsky received a worldwide patent for two pharmaceutical products: Rupununine, extracted from seeds of the *Octotea rodioei*, for birth control; and Cunaniol, a nervous system stimulant, extracted from *Clibadium sylvestre*. The use of the plants is part of the traditions of the native Wapixana Indians, who live in the Brazilian state of Roraima. Several other bioprospecting projects are underway in Africa and other places to identify plant extracts, animal toxins, and microorganisms for different purposes such as production of plastics or ore purification and fermentation processes.

When a sample of a species is collected illegally and a new drug or an isolated gene from that sample is patented, the patent can be revoked. If there is proof that the active ingredient used in the new drug was in public use, even if restricted to an indigenous tribe, revocation of the patent is possible. The great dilemma in patenting a natural product is that pharmaceutical companies take advantage of the ethnobiological knowledge of indigenous populations, and later, the companies are the only ones to collect profits from the marketing and production of the drugs.

To prevent such exploitation, regulations are being made worldwide to govern the use of biological diversity. In 1992, the United Nations Conference on Environment and Development (ECO 92) met in Rio de Janeiro, Brazil, with representatives from 120 countries. This conference recognized the national sovereignty of the nations and the genetic resources within their borders. Beyond this international work, the national laws of each country further govern the conservation and development of biodiversity within their respective boundaries. The Brazilian Congress recently recognized the importance of the protection of its biodiversity. This came after an accord with some multinational companies relating to the development of medicines resulting from the exploration of plants and microorganisms from the Atlantic rainforests and the Amazon.

Only 20 years ago, legal aspects related to the collection of samples of plants, microbes, and animals were largely ignored. In most cases, researchers simply made a trip to the place where the species of interest could exist in nature, collected the samples, and returned to their laboratories. Clearly laws didn't exist to regulate that practice. Sometimes, researchers obtained informal authorization from the local authority or from the landlord where the samples were collected. The days of locating, collecting, and returning home are running out, at least legally in most countries. More often it is today considered biopiracy.

FINAL CONSIDERATIONS . . . .

Legal mechanisms should be developed to protect the biodiversity of the world. The mechanisms should promote the conservation of biodiversity and its use for the well-being of mankind. Developed countries and many others have taken the lead to ensure germplasm preservation for years to come. Extensive efforts are being made to characterize, catalog, and store the germplasm resources collected over many years from all over the world. Further steps are being taken to guarantee the preservation of the biodiversity of ecosystems and centers of origin for future research. Biotechnology will benefit from the world's biodiversity, while creating a means of preservation and continuation of the diversity of life found around the world.

There is an incredible amount of biodiversity worldwide, and much of it is relatively unknown. Despite the actions to explore this diversity, steps are being taken to characterize and preserve this valuable resource.

13 Bioterrorism

*B*ioterrorism is the term used to descibe the offensive employment of biological substances or toxins with the objective of causing harm to an individual or a group of individuals. These activities, in general, cause damage, intimidation, or coercion, and are usually associated with threats causing public panic. The most common biological agents used as weapons are microorganisms and their associated toxins, which can be used to promote disease or death in people, animals, and even plants. The agents of contamination can be dispersed in the air, water, food, and elsewhere.

Bioterrorism has been a problem throughout human history. One of the first reports of bioterrorism dates back to the 6th century B.C., when the Assyrians poisoned the wells of their enemies with ergot, a toxin-producing fungus often found in rye. A more recent report suggests that in the 1500s, Pizarro, in the conquest of South America, gave clothes contaminated with smallpox to native Indians. Another similar report alleges that Britain might have used pathogens to weaken their opponents during the colonization of North America. The country might have deliberately distributed blankets contaminated with smallpox to Native Americans. Terrorism using chemical or biological weapons often spreads quietly, but it can have devastating impacts.

The first convention banning biological weapons was signed in Geneva in 1925. In 1972, under United Nations leadership, 103 countries signed the Convention on Biological Weapons, which prohibits the development, production, stockpiling, and use of biological weapons. The objective of this convention was to completely eliminate the use of biological agents and toxins as weapons of mass destruction.

During a conference on bioterrorism held in San Diego, California in early 2000, experts concluded that the United States was not prepared for a biological attack with pathogens such as smallpox, anthrax, Ebola, botulism, and others. At the second National Symposium on Bioterrorism in Washington, DC in 2000, one of the conclusions was that the American public health system was not prepared to respond to an attack with biological weapons. Additionally, in March 2001, researchers at the Center for the Study of Bioterrorism and Emerging Infections at the St. Louis University School of Public Health revealed that 75 percent of health agents feared that some city in the United States would suffer an attack with biological weapons within the next 5 years. The forecasts by the experts were correct: In October 2001, just four months after the meeting, the United States had its anthrax attack.

Anthrax and especially smallpox are considered the most serious threats of biological bioterrorism, due to the high fatality rate among infected individuals, the possibility of transmission in an aerosol form, and the relative ease of large-scale production. Various government-sponsored biological warfare programs have researched many other pathogens as well (Table 13-1).

Table 13-1
Biological Warfare Programs in Different Countries

Country	Status	Period	Disease	Observation
Canada	Stopped	1941–1960?	Anthrax, bovine pest	Exact date of termination not known
Egypt	Stopped	1972–present	Anthrax, brucelosis, mormo (*Malleomyces mallei*), psittacoses, equine encephalitis	
France	Stopped	1939–1972?	Potato beetle, bovine pest	Exact date of termination not known
Germany	Stopped	1942–1945	Anthrax, foot and mouth disease, mormo (*Malleomyces mallei*), potato beetle	During World War II also used other agents
Iraq	Active	1980–present	Aflatoxin, anthrax, camel smallpox, foot and mouth disease, wheat rust	Suspected to still have an on-going program
Japan	Stopped	1937–1945	Anthrax, mormo (*Malleomyces mallei*)	During World War II also used other agents
North Korea	Active	?–present	Anthrax	
Rhodesia	Stopped	1978–1980	Anthrax	An anthrax epidemic resulted in 182 human deaths
Syria	Active	?–present	Anthrax	

(*continued*)

Table 13-1
Biological Warfare Programs in Different Countries (*Continued*)

Country	Status	Period	Disease	Observation
United Kingdom	Stopped	1937–1960?	Anthrax	Exact date of termination not known
United States	Stopped	1943–1969	Anthrax, brucelosis, equine encephalitis, foot and mouth disease, mormo, potato blight, new castle disease, psittacoses, rice blight, bovine pest, wheat rust	
Soviet Union	Stopped	1935–1992	Anthrax, African swine fever, poultry flu, brucelosis, contagious bovine pneumonia, contagious ectima, foot and mouth disease, mormo (*Malleomyces mallei*), corn rust, new castle diseases, potato virus, psittacoses, bovine pest, rice blight, TMV, Venezuelan equine encephalitis, wheat and barley viruses	Also researched with insects and other agents

The list of pathogens with potential terrorist applications ranges from salmonella to supervirulent (highly infective) strains of the bacteria that causes bubonic plague (*Yersinia pestis*) genetically modified by recombinant DNA technology. There are also toxins such as ricin, the organic phosphorous sarin gas, or the Ebola virus. Recently, the North Atlantic Treaty Organization (NATO) listed 39 biological agents that could be used as weapons by terrorists.

A wake-up call for the risk of bioterrorism occurred when sarin, a gas affecting the nervous system, was used in an attack carried out by the religious cult Aum Shinrikyo in a Tokyo subway in 1995. Interestingly, according to the Monterey Institute of International Studies (*http://www.miis.edu*), of more than 100 other terrorism acts recorded since 1960, the great majority has failed. However, interest in bioterrorism has increased significantly after the anthrax cases that occurred in late 2001. The U. S. Department of Health spent, as of the last decade, about $160 million annually on bioterrorism prevention. After the September 11 terrorist attack and the anthrax cases that followed, it is believed that investments in this area will rise substantially with an increase in the reality of terrorist threats.

As discussed, biotechnology can be used in the development of pathogens with higher virulence and increased antibiotic resistance, when in the wrong hands, but the science can also be used in the development of biodefenses as well. Early warning indicators, more precise diagnostic procedures, therapy, vaccines, pathogen identification, and new pharmaceuticals are only some of the areas in which biotechnology can be of help in the area of bioterrorism.

The main objectives in preventive bioterrorism are the production and stockpiling of vaccines and the development of early warning systems in the event of an attack. Genome sequencing also promises to facilitate the development of biodefenses and decontamination. An ongoing project at the University of Michigan is developing a mechanism to kill anthrax, using a solution of droplets of soybean oil in aqueous suspension. The droplets of this emulsion fuse with the bacterial membrane by means of a chemical reaction, thereby generating sufficient energy to destroy bacterial spores. Another interesting area of research is the production of synthetic antibodies to improve the treatment of infection.

Experts know that most of the progress in biotechnology is not only useful to combat biological agents spread intentionally, but also for naturally occurring disease epidemics. Considering that the real threat of bioterrorism is present worldwide, preventative measures are catching the attention of governments and of the public.

The use of bioweapons requires the cultivation, purification, stabilization, and large-scale production of the pathogen, as well as the development of an efficient means for dispersal. For instance, the dispersion of bacterial spores with a size ideal for uptake into the bronchioles of the lung can be an additional challenge for terrorists without scientific knowledge. Experts believe, however, that it is not difficult to find leads to bioterrorism programs in the international market. It is widely believed that thousands of trained scientists with expertise in biological warfare lost their jobs after the collapse of the former Soviet Union in December 1991.

HUMAN PATHOGENS OF POTENTIAL USE
IN BIOTERRORISM . . . .

The most dangerous bioweapons are contagious organisms with tolerance to extreme climatic variations and a long incubation period. This could include human viruses, bacteria, and fungi. Although the epidemic risk is significant with some of these microorganisms, past attempts to use biological agents have had limited success, mainly due to the complexity of the interaction of these microorganisms with the environment. Air humidity and temperature do not always favor the development and growth of the pathogen. The following sections cover pathogens with the potential to be involved in bioterrorism.

Anthrax

The bacterium *Bacillus anthracis* is an infectious and deadly pathogen, but it is not contagious. This bacterium can be easily found in farms, where it infects cattle and other animals. Children and adults can show three clinical forms of the disease: cutaneous (Figure 13-1), inhalation, or gastrointestinal. The symptoms and signs of anthrax in children and adults are similar, and they usually are seen about

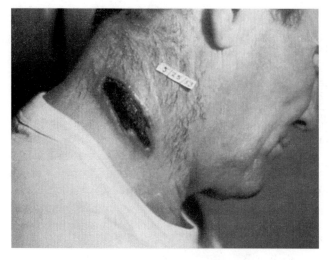

Figure 13-1
Symptom of cutaneous anthrax.

seven days after the infection. Because symptoms mimic those of the flu, the correct clinical diagnosis could be difficult, requiring specific laboratory analyses.

Most scientists agree that only individuals with access to advanced biological resources can produce anthrax for aerosol dispersion. The literature reports that an accidental dispersion of *Bacillus anthracis* spores in 1979 in the former Soviet Union caused the death of 68 people. The U.S. Centers for Disease Control and Prevention (CDC), the governmental agency in charge of monitoring public health, recorded 18 confirmed cases of anthrax in 2001, including seven cutaneous infections and 11 cases of inhalation infections, five which resulted in death. These cases were associated with mail contaminated with spores of the pathogen, sent to different governmental and television personalities (Figure 13-2). The FBI has not ruled out the possibility that international terrorists or domestic fanatics sponsored these events. This agency is also following up on the possibility that the anthrax cases could be associated with individuals who might have profited from the attacks.

Smallpox

Smallpox is a contagious viral disease that has been virtually eradicated from the world since 1980. Since that time vaccinations have nearly ceased around the world, placing the entire population at risk

Figure 13-2
Envelope contaminated with spores of *Bacillus anthracis* sent to U.S. Senate majority leader Tom Daschle in October 2001.

Source: From Federal Bureau of Investigation.

of epidemics. This single disease caused the death of 300 million people during the 20th century. A small stock of this virus was maintained in the United States and in Russia for biomedical research. According to the World Health Organization (WHO), it was supposed to be destroyed in 2002. Scientists believe that smallpox represents the single largest risk of human mortality in case of an outbreak of the disease. In the United States, vaccination for smallpox ceased in 1972. Therefore, the current American population under the age of 30 doesn't have the benefit of the vaccination at all. An enormous portion of the population would thus be susceptible to the disease.

Currently there are about 12 million doses of smallpox vaccines in the United States, which is enough for less than 5 percent of the population. However, it is expected that by the end of 2002 government will have approximately 280 million doses, produced through the benefits of biotechnology.

Cholera

Vibrio choleare, the bacterium that causes cholera, was endemic in many parts of the world until a few decades ago. This bacterium is frequently found in untreated water. Usually, this disease is not transmitted from infected patients.

Salmonella

In 1997, the Rajneeshee cult in Oregon contaminated salads with *Salmonella enteritis* with the goal of affecting the results of the presidential election. However, this bacterium is not considered dangerous enough to be used as a serious weapon by terrorists. Contaminated food can be quickly removed from the market, and the mortality rate is low among patients.

Botulism

There are different avenues of infection by the bacterium *Clostridium botulinum*: through food, in intestinal ulcers, or by inhalation. This inhalation form causes facial paralysis after the third day of infection.

An effective treatment depends on early diagnosis and the administration of an antidote. If undiagnosed, the botulism toxin produced by the bacteria can be fatal.

Poliomyelitis

Most developed countries have eradicated poliomyelitis, the disease also known as polio. However, some countries in Africa still struggle to wipe out this disease. The WHO estimates that polio will be eradicated from the world by 2005.

Ebola

The Ebola virus was first detected in 1976 in the Democratic Republic of Congo (formerly Zaire) near the Ebola River. This virus destroys the human immune system, causing death in about 90 percent of those infected. Unfortunately, there is no known cure for this disease. The principal symptoms are fever, chills, anorexia, nausea, exhaustion, hemorrhagic conjunctivitis, and widespread hemorrhaging. Recently, two epidemics were identified in Gabon and in the Democratic Republic of Congo.

TOXINS . . . .

For hundreds of years Indians have added bacteria, fungi, algae, and plant and animal toxins and venoms to weapons to increase their efficacy. These substances started to draw the attention of military forces before World War II. However, the military soon lost interest because of the low stability of these substances and the difficulty of large-scale use. However, recombinant DNA technology, when improperly used, could be applied to modify genes that code for these toxins, so that the recombinant toxin is more stable, more potent, and consequently more destructive. Although toxins are not considered substances for large-scale use, they have been used in some isolated incidents and in a few common crimes. These toxins are listed in Table 13-2.

Table 13-2
Biological Agents Used as Weapons

Agent	Traditionally Used in Warfare	Used in Bioterrorism and Common Crimes	
Pathogen	*Bacillus anthracis*	*Ascaris suum*	*Salmonella typhi*
	Brucella suis	*Bacillus anthracis*	*Vibrio cholerae*
	Coxiella burnelli	*Coxiella burnelli*	Ebola virus
	Francisella turalensis	*Giardia lambia*	Yellow fever virus
	Smallpox virus	*Rickettsia prowazekii*	*Yersinia enterocolica*
	Viral encephalitis	*Salmonella typhimurium*	*Yersinia pestis*
	Ebola virus		
	Yersinia pestis		
Toxin	Botulin	Botulin	Snake venom
	Ricin	Cholera	Tetradotoxin
	Staphylococcal	Ricin	

Botulism

Produced by *Clostridium botulinum* (mentioned previously), this is the most dangerous known toxin. *C. botulinum* usually grows in spoiled canned food. The incubation period of botulism is from one to three days, after which the victim manifests stomach aches, diarrhea, vision problems, and muscular fatigue. Eventually, the respiratory system becomes paralyzed, resulting in suffocation and death a few days after contamination. The toxin exists in seven different forms. The LD_{50} (or lethal dose for 50 percent of the population) for an average man is estimated to be less than 1 μg if ingested, and even less if inhaled. Just 1 g of botulism dispersed in the air can kill 1 million people. Vaccines for botulism are available, and they are the best form of protection for the population. But once an individual has inhaled the toxin, an effective antidote does not yet exist. The antidote is effective only in cases of ingestion, and it should be taken immediately after contamination. Despite its toxicity, forms of the botulism toxin have been produced for treatment of some muscular disorders.

Ricin

This is a toxin present in the seeds of the castor bean (*Ricinus communis*). In 1978, it was used in the case of the "Umbrella Murder" in London, when it was used in a bullet to kill a fugitive from Bulgaria. Ricin inhibits protein synthesis in the human body. Through genetic engineering, ricin has also been produced in *Escherichia coli* transformed with the toxin gene. This is being used in the investigation of liver cancer therapies.

Trichothecenes

These fungal toxins are produced by *Fusarium* species of fungus. It is believed that these were used in the 1980s in Southeast Asia under the name yellow rain.

Staphylococcal Enterotoxin

Produced by the bacteria *Staphylococcus aureus*, this is the most common substance associated with food poisoning. The toxin is water soluble and relatively thermally stable, withstanding boiling. Individuals exposed to 20 to 25 g of this toxin manifest colic, nausea, and diarrhea within a few hours after ingestion.

Saxitoxin

This small toxic molecule is produced by cyan-blue algae. It attacks the nervous system and can cause paralysis. LD_{50} for humans is 1 mg. Marine scientists have observed mass deaths of humpback whales and other sea life caused by saxitoxin. Although the cause of death has been initially blamed on pollution, as none of the animals showed any signs of disease following postmortem examination, death was later associated with saxitoxin poisoning. Symptoms include dizziness, diarrhea, vomiting, disorientation, respiratory distress, and eye irritation.

BIODEFENSES . . . .

Recombinant Vaccines

Sequencing the genome of pathogens allows the development of more efficient vaccines. In April 2000, the Pasteur Institute in France sequenced the genome of the bacterium *Mycobacterium leprae,* the causal agent of leprosy. The genomic sequence was also obtained for many pathogens, including *Plasmodium falciparum* (malaria), *Corynebacterium diphteriae* (diphtheria), *Neisseria meningitidis* (meningitis), and *Enterococcus faecium* (an antibiotic-resistant infectious bacteria), promising to accelerate the development of new weapons against these diseases. The knowledge of the genomic sequence of the pathogens allows scientists to identify their weaknesses, making it possible to develop not only vaccines but also more efficient therapies.

Edible Vaccines

The production of vaccines in recombinant microbes, animals, or plants has been drawing great interest from public health agencies. The WHO has been recommending the use of vaccines as one of the most efficient strategies for disease prevention. Although vaccine research represents only 3 percent of total pharmacological research and development funds, it is considered to possess an excellent cost–benefit ratio.

Plants have been considered an excellent means for the production of recombinant vaccines because they are easily propagated and the vaccines can be expressed and in seeds or fruits. This method bypasses any risks of contamination from animal pathogens such as viruses that can also infect humans. Plant tissue expressing the vaccine can be used directly for human consumption, as an edible vaccine. This alternative would also eliminate the need for purification and refrigeration, which is currently required for most vaccines.

Research with edible vaccines has been focused on gastrointestinal diseases, caused by *Escherichia coli*, *Vibrio cholerae*, Norwalk virus, and rotavirus. Hepatitis B, Type I diabetes, and autoimmune diseases have also been subjects of investigation. One of the main aspects still to be understood with edible vaccines is how to gauge the correct dosage of plant tissue that should be ingested to provide sufficient immunity. It is important to recognize that the transgenic varieties in development for production of vaccines, antibodies, or any other pharmaceutical should be considered a medicine and not a food for general use.

Biotechnology can help in the immunization of the population and the consequent eradication of diseases with transgenic edible vaccines. Banana and potato plants genetically modified to express vaccines are being developed in different institutions. These vaccines can have an important role in immunization against disease because they require no special storage conditions and they could be grown and processed in the area in which they are needed. One of the inconveniences with these products is that they should not be cooked before consumption, because cooking tends to break down the vaccine.

AGROTERRORISM ....

Until September 11, 2001, the date of the terrorist attacks on New York City and Washington, DC, the fear of terrorism was limited in the Western world and its risk was associated primarily with airliners. More recently, the government is considering all forms of terrorism. Some of the possible targets of terrorism include nuclear plants, large buildings, national monuments, water reservoirs, and the food supply. This brings up the threat of agroterrorism. *Agroterrorism* can be defined as the deliberate and malicious use of biological or chemical agents as weapons against agriculture, and more specifically, the food supply. The destruction of the food supply or its contamination with noxious agents can disrupt people and nations.

The safety of the food industry, including farming, processing, and distribution should be considered a matter of national security. Even so, the food industry is vulnerable to sabotage in nearly all steps of the production chain (Figure 13-3). Because of this, most scientists agree that the risk of an agroterrorism act cannot be precisely measured.

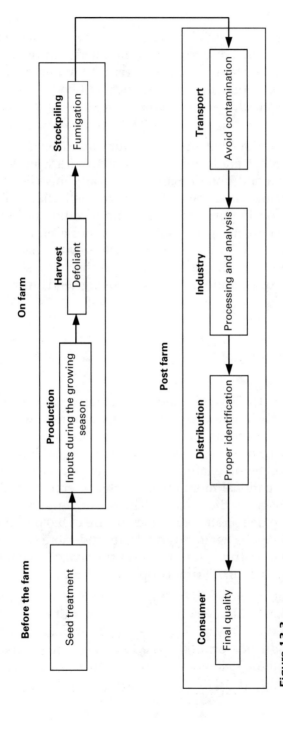

Figure 13-3
Flowchart of the food industry.

Recent epidemics, such as the case of mad cow disease in Britain and of foot and mouth disease in different countries, supply important empirical data for the simulation of impacts from an agroterrorist attack. The mad cow epidemics revealed that the economic impact of this disease reached not only to the agriculture industry, but also to tourism and other industries, and the effects continued outside the borders of the epidemic. It is believed that an agroterrorist attack could have a much larger impact than natural epidemics, because they are deliberately planned to do harm. Experts suggest that the damage from an attack would depend on the time required to diagnose the problem. An early warning system might be as important as the means to counter the attacks. As the food production chain is extremely complex, with several steps (Figure 13-3), there are many vulnerable points that could disrupt the food industry.

Some of the catastrophes that have happened in agriculture might shed some light on the complexity and fragility of this industry. In 1970, the plant pathogen *Bipolaris maydis (Helminthosporium maydis;* Figure 13-4) destroyed hybrid corn fields across the United States. This disease, resulting in damage of about $1 billion, affected all corn

Figure 13-4
Healthy corn leaf (left) and one infected with *Bipolaris maydis* (right).
Source: Courtesy of USDA-ARS, photo by Keith Weller.

hybrids possessing T cytoplasm. In Belgium, the contamination of beef, dairy products, milk, chicken, pork, fish, and eggs with dioxin, a cancerous substance, resulted in a ban of Belgian produce around the world. The economic loss to Belgium was estimated to be approximately $1 billion.

Several cases of agroterrorism are also worth mentioning:

1. The 1974 attack by the group Revolutionary Palestinian Commando, which allegedly contaminated a shipment of Israeli grapefruits to Italy.
2. The 1978 attack by the Revolutionary Palestinian Council on citrus orchards in Israel using mercury.
3. In 1999 and 2000, eggs produced in Israel were contaminated with salmonella. In this last attack, two people died and many were hospitalized.

There are suspicions that in 1989, a radical group released Mediterranean fruit flies in California to protest the use of insecticides. This insect caused serious financial loss to Californian citrus farmers that year, but no one was charged. There are suspicions that the cotton Boll weevil (*Anthonomus grandis*) was intentionally introduced into Brazil in the 1980s to interfere with national cotton production. When economic and political interests are involved, one cannot doubt that unscrupulous groups could be willing to use these strategies.

During World War II, the United States, the Soviet Union, Japan, Britain, and Canada studied several animal and plant pests with warfare objectives. Anthrax, brucelosis, foot and mouth disease, equine encephalitis, wheat rust, and several other pests were studied (see Table 13-1). Although the reports in the literature are contradictory, there is evidence suggesting that Japan used animal and plant pathogens against the former Soviet Union during World War II. In 1972 the United States, Soviet Union, England, and Canada signed the Biological and Toxic Weapons Convention, an international treaty to ban an entire class of bioweapons. Some countries, such as Iraq, have not yet signed on to this convention.

The low technological level and simplicity of the equipment required for an agroterrorism program make underground activities extremely difficult to detect and monitor. A technician with limited training in microbiology can isolate and multiply pathogenic microorganisms in a backyard laboratory. However, an efficient attack requires the acquisition, multiplication, and processing of a biological agent. Furthermore, it requires the development of an efficient distribution mechanism and techniques to account for unfavorable environmental conditions (air humidity, temperature, winds, etc.). The failure of most of the previous attacks with biological agents has been attributed to inadequate weather conditions for the released agent to infect its host, multiply, and start secondary cycles of the disease. Although most countries have some type of restriction for the introduction of pathogens, it is possible, under the guise of research, to acquire many of these biological agents from several international laboratories.

RECOMMENDATIONS . . . .

Although it is impossible to predict where, when, or what kind of bioterrorism poses the next threat, many scientific societies have come up with recommendations for the government, including the following:

- Sequence the biological pathogens that are a risk to be used as a bioweapon. This is useful to determine successful strategies to combat the effects of the pathogen.
- Stimulate functional genomics research to understand virulence, pathogenicity, and biology of pathogens.
- Develop tools for rapid, accurate diagnosis.
- Use broad-based, durable resistance in new crop varieties.
- Improve the understanding of host–pathogen interaction at a molecular level.
- Perform surveillance of pest outbreaks.
- Increase research on pathogens for peaceful purposes.
- Establish a real-time disease reporting system.
- Develop a list of facilities and experts who are familiar with the pathogens.

- Establish a biosafety committee.
- Implement training programs for front-line responders and detectors.
- Develop a network of diagnostic laboratories.
- Develop a national center for agricultural control.
- Increase efforts in global surveillance of emerging and high-risk diseases.

FINAL CONSIDERATIONS . . . .

Although pathogens can be genetically modified to increase virulence and the capability of promoting an epidemic, there is no evidence that this has already occurred. However, this is no guarantee that the world is free from this risk.

As bioterrorism is a matter of national security, it is the responsibility of intelligence agencies to ascertain its real extension and potential. Some military experts believe that Iraq still possesses an active biowarfare program. A rare disease caused an epidemic in Iraqi wheat fields some years ago, arousing suspicion that a virulent pathogen had escaped from laboratories conducting bioterrorism research. Any preventive program in bioterrorism should involve intelligence, constant monitoring, early warning systems, information sharing among agencies, and cooperation among others. There should be laws in place that would allow the government to enforce quarantines of suspected infected individuals or goods, confiscation of properties, and use of hospitals to provide for the common good.

Finally, no one should assume that biology and the science of biotechnology will always be used for good. In states that sponsor terrorism, biotechnology could be used to develop pathogens and pests for mass destruction. Recent events have awakened our awareness to the global community in which we all live, and local events often have worldwide impact. It is important to be aware that the science of biotechnology, with all of its benefits, could also advance the unfortunate efforts of terrorism.

For more information on bioterrorism, visit the following Web sites:

- American Biological Safety Association: *http://www.absa.org*
- Biological Warfare Agents:
  http://www.emedicine.com/emerg/topic853.htm
- Centers for Disease Control and Prevention:
  http://www.bt.cdc.gov
- Department of Peace Studies, University of Bradford:
  http://www.brad.ac.uk/acad/sbtwc
- International Security Information Service:
  http://www.isisuk.demon.co.uk
- Center for the Study of Bioterrorism and Emerging Infections:
  http://www.slu.edu/colleges/sph/bioterrorism

14 Bioethics

Ethics, in general, deals with matters related to moral concepts and behavior patterns that are socially or morally sound and acceptable. Concepts of ethics differ from country to country according to culture and traditions. They also change with time, due to shifting perceptions of values, which are deeply affected by scientific and technological progress, as well as the media and other forms of popular information. Bioethics seeks to study the moral vision, decisions, and politics of human behavior in relation to biological phenomena or events.

Ethics not only deals with life (e.g., in vitro fertilization, prenatal genetic selection, cloning, sperm banks, gene manipulation, and gene therapy), but also relates to death (e.g., euthanasia, maintenance of those in comatose states). In this chapter, only ethical aspects related to biotechnology are discussed.

There is no other issue more controversial and frightening than bioethics. It is an area with a hand in science and another in public values and beliefs. Until some decades ago, specialists in bioethics had, in most cases, only a formal education in philosophy or theology. Today, the reality is quite different. In the United States, there are about 20 formal degree programs in bioethics available at different universities. This derives from the demand for professionals in bioethics, which has risen in the last 20 years, a period when human, animal, and plant genetics has ventured into avenues never before imagined.

Biotechnology has drawn a wide range of reactions in society, mainly based on the individual's own opinions and perceptions, and not facts or validated information. An individual's perception of controversial issues, such as biotechnology, is usually based on his or her background and moral or religious beliefs. Therefore, different communities, ethnic groups, and cultures have quite varied ethical values. Additionally, recognizing the fact that society is dynamic and co-evolves with scientific progress (inventions, discoveries, etc.) helps one realize that ethics is also dynamic. In addition, every group of people might react differently to a certain issue. For example, the speed with which new technologies is adopted is quite striking. In general, the human population can be subdivided into three subgroups:

1. Those enthusiastic about new technology, including individuals ready to adopt new technologies. These tend to be the most innovative people.

2. Those against any new technology. These people refuse to accept new technologies, and they are generally the most conservative.

3. The progressives, those who offer initial resistance to new technologies, but are eventually educated about the technology and adopt it.

The era of biotechnology has brought a wide spectrum of new topics with which most people do not feel comfortable offering their own opinions. Some of the common topics include transgenics, genetic tests, prenatal selection, gene therapy, cloning, genetic discrimination, and eugenics. These subjects are addressed here with the objective of assisting you in developing a critical perspective about them. For some, biotechnology means promise, but for others, it is a reason for concern. The possibility of a cure for genetic diseases such as cystic fibrosis, breast cancer, Alzheimer's disease, Huntington's disease, and many other devastating ailments is of value to patients and their relatives. These promises have an even greater tangible value when they affect those closest to us. However, as industry and big business come into play in this scenario and use information generated by biotechnology to increase their revenue, ethical concerns are obvious, creating the need to make ethical decisions relating to this new technology.

ETHICS AND GENETIC ENGINEERING . . . .

Science has proven that DNA is the basis of heredity in nearly all living creatures. It is unusual to think that the same molecules that make a fungus a living creature are also similar for human life. The science of genetics has even found gene sequence coding for specific enzymes and proteins that are virtually identical in humans, plants, and microorganisms. Experiments have shown that genes from one species can be manipulated and expressed in another species. This forms the basis of the science that has been discussed in the previous chapters.

With the advances from biotechnology in agriculture, medicine, and other areas, genes from highly diverse organisms have been transformed into other species to obtain the expression of a certain trait. An often-cited example is again that of Bt corn. A gene from a soil-borne bacterium was engineered into corn to provide resistance to a devastating insect species. For many, this is not ethically wrong, but others find inherent problems in the use of genes across species. It is a basic issue with the science of biotechnology: whether or not our knowledge of DNA and genetics should allow us to manipulate organisms that are not naturally compatible. Many believe that such genetic manipulation is beyond the realms of responsible and moral science. Does such DNA manipulation change the inherent properties of corn or any other organism, or does biotechnology serve to expand the frontiers of life?

This question is one of the basic ethical arguments behind biotechnology and is actually just the beginning of the many ethical questions that can be directed at this science. Despite arguments about the moral justifications of basic genetic engineering, scientists continue to develop products using advanced methods of gene manipulation. However, many cultures and traditions might be affected by such engineering techniques. For instance, those of the Jewish faith abstain from the use of pork, as it is traditionally considered unclean. What would be the ethical implications of using swine genes in a medicine, plant, or other product? Would it compromise the faith of one who abstains from pork? Such examples can be expanded to include many other scenarios in which this encounter of science with tradition could occur. The issue leads to the questioning of many long-held traditions and beliefs. Perhaps life is simpler than previously thought, and advances in genetics and biotechnology allow us to understand how life is simply contained in the ordered chemistry of DNA. This leads to the importance of public awareness of the applications of biotechnology and must also be included in a debate about ethical implications related to this expanding science.

GENETIC PRIVACY . . . .

Privacy and confidentiality are one of our most valued possessions. Soon, it might not be possible to hide from society our weaknesses, limitations, and genetic deficiencies. The individuality of each human being is being unmasked.

The completion of the first version of the Human Genome Project, announced in February 2001, unveiled the genomic sequence of the almost 3.2 billion letters of our chromosomes. This announcement also generated discomfort and fear. The revelations of the deficiencies and predispositions coded by human gene sequences are an unsettling idea for many people. The knowledge of the human genome sequence will make possible the diagnosis of several diseases even before their onset. The association between genes and genetic diseases not only affects patients, but it also raises legal and economic issues for society.

In some countries, like Great Britain, the population of some regions has voluntarily donated DNA samples for establishing a genomic database for criminal use. In that country, the ethical perception is that DNA donation for genomic databases is not an invasion of an individual's privacy. In the United States, the situation is completely the opposite: The great majority of the population has a much stronger sense of individuality and so opposes the idea of genetically exposing themselves. However, some genomic databases have been made mandatory in the United States, such as the one at the FBI. In some states, inmates involved in sexual crimes, murders, and other violent crimes have been required to have their DNA profile included in genomics databases. The fear of many people in relation to the loss of their genetic privacy is that the DNA information could be used for discrimination in employment situations or for health and life insurance.

Health insurance companies in some countries have begun to use two tables for deciding premiums: one for the carriers of genes associated with colon cancer and another for individuals that do not have those genes. The idea of pre-existing conditions as a limiting factor in insurance coverage could soon be extended not only to diseases manifested previously, but also to the presence of genes associated with predisposition to diseases or weaknesses.

Some geneticists believe that a DNA profile will not only be used to solve crimes, but also to prevent them. They argue that violence is a genetically transmitted trait. Dutch scientists studied a family in the Netherlands with male individuals with outstanding rates of aggressiveness and rape over five generations. The men were found to carry a genetic defect that causes a deficiency of the enzyme serotonin in their brains. Serotonin is a neurotransmitter related to behavior, humor, and personality in humans. Perhaps this is evidence for the existence of genes associated with violence and other antisocial behavior.

A correlation has also been found between atypical levels of serotonin and dopamine with violent behavior and suicide in different studies. The levels of the neurotransmitter are regulated by genetic and environmental factors as well as the interaction between the two factors. The environment is a broad term referring to nongenetic factors contributing to the trait of interest. This can include upbringing, relationships, lifestyle, and other less tangible influences on behavior or health. Current attempts to link violence to genes are more sophisticated and are applied to individual cases and not to population groups. It is believed that, in the near future, genes for violence or dishonesty will be identified, just like the genes related to cancer and other diseases. In reality, even if all the complex human traits like intelligence, violence, honesty, anxiety, and friendliness are mapped and sequenced, they will remain misunderstood for a long time because the understanding of the human mind is still so limited.

Even if the existence of a gene for violence were confirmed, would it be ethically correct to label a child as prone to violence? Would the simple knowledge of that information alter the isolation patterns of society in relation to a carrier of that gene, inducing him or her to become violent? It should be recognized that the manifestation of most traits results not only from genes, but also environmental influence during an individual's life.

"Family Sues State: Son's Propensity to Violence Does Not Justify Discrimination." That type of headline has not yet been printed, but it might be a nearer reality than you imagine. The discovery that genes A, B, or C are associated with violence or dishonesty will engender ethical issues with a profound impact in modern society. How would society accept the use of genetic information to prevent crimes and not only solve them? Many psychologists already support the hypothesis that certain genes code for violence.

The idea that all behavior, tendencies, health, and other traits are controlled by genes is called *gene myth*. Despite statistical or other associations between genes and some traits, a great part of behavior and health is determined by lifestyle, relationships, and upbringing.

There might be genes related to lung cancer, but that doesn't discount the issue of smoking as a major factor for the disease. Likewise, media and the examples of friends and family heavily influence genetic tendencies toward violence or honesty. Many believe that a greater understanding of our genomes will result in scientific explanations, or excuses, for antisocial or criminal tendencies, when there are likely other, more important reasons for such traits.

The DNA profile of an individual not only reveals his or her propensities to develop diseases but also his or her intrinsic potential. Universities, corporations, and insurance companies, among other institutions, will certainly have an interest in accessing information from genomic databases before offering employment or an opening to courses to customers.

Less complex traits like height can be studied and understood more objectively. About 5 percent to 10 percent of 200,000 people with breast cancer diagnosed annually are found to possess genetic factors associated with the disease. The genes most commonly associated with breast cancer today are BRCA1 and BRCA2 (Figure 14-1),

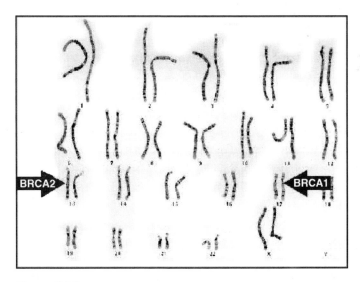

Figure 14-1
BRCA1 and BRCA2 genes are located on human chromosomes 13 and 17.
Source: Courtesy of Myriad Genetics.

two different mutations in the normal form of genes. BRCA1 and BRCA2 also increase the risk of ovarian cancer in women and prostate cancer in men.

In families with a history of breast cancer, especially in those of Jewish descent from Eastern Europe, the incidence of the mutations BRCA1 and BRCA2 are much higher than in the general population. In such situations, women should be tested for the two genes. If the result is positive, the chance of developing the disease is greater than that of the normal population. Although the presence of the mutated gene is not a guarantee of cancer, some women still opt for a radical mastectomy as a preventative measure.

The knowledge of the presence of the BRCA1 and BRCA2 genes in individuals can lead them to alter their lifestyle, reducing consumption of alcohol and increasing the frequency of mammograms. In such cases, genetic tests can alter the individual's fate.

For other diseases the situation could be much more complex. For example, Huntington's disease is a degenerative and fatal neurological disease that usually has an onset around the age of 45 to 50 years, an age at which most people have had children. In families in which one of the parents is affected by this disease, each child has a 50 percent chance of also having the disease. People that have parents with Huntington's disease tend to not have genetic tests; they prefer to live with uncertainty about the disease rather than the certainty of developing the disease, because there is currently no cure, and there are few treatment options.

Today there are genetic tests for the detection of genes that predispose an individual to the following diseases: sickle cell anemia, Down's syndrome, Huntington's disease, muscular dystrophy, cystic fibrosis, Tay-Sachs, colon cancer, breast cancer, Alzheimer's disease, and multiple sclerosis. The number of diseases for which genetic tests are available continues to grow.

The inclusion of genetic tests results in information on an individual's medical record that could have serious effects in his or her life. Health or life insurance companies might refuse coverage for medical treatments under the allegation of a pre-existing condition. Today, pre-existing conditions only apply to diseases that have already been manifest in the individual; in the future, this might be extended to include the presence of genetic factors linked to specific medical conditions.

Large corporations routinely request intelligence and personality tests for prospective employees. Some people fear that, in the near future, genetic tests will be routinely requested prior to employment. Today, some companies already use genetic tests to identify employees who are sensitive to chemical products used in the work environment. Those companies have argued that the genetic tests are used only to protect their employees from risks related to work, and obviously, to eliminate the risk of being sued for damages in the future. According to the companies, they do not use genetic tests for selection purposes. In the future, companies could opt to increase the number of genetic tests that are mandatory for recruitment. Such tests might reveal personality patterns and could possibly be used for discriminatory means.

Biotechnology has opened the door to our lives, and questions of genetic privacy still remain to be answered. It is now possible to know more about our genetic makeup, but is that necessarily good? This raises many questions about the use of the enormous amount of information that has been made available.

PATENT OF GENES . . . .

John Moore, a leukemia patient, was treated at the University of California Hospital in Los Angeles in 1980. As part of his treatment, his doctor removed part of his bone marrow. With cutting-edge treatment, his leukemia went into remission, but during the following years Moore continued to have samples of bone marrow, blood, epidermis, and semen taken. Eventually, he became uncomfortable with the endless sampling of his tissue and decided to seek legal advice. To the surprise of many, he discovered in 1984 that the University of California had patented one of his cell lines.

The Patent Act was originally issued to protect "any art, machine or material that is new and useful." Until some decades ago, the thought of patenting a living organism was inconceivable. However, when a patent for a bacterium with the capacity to decompose petroleum was requested, a new era of patenting began. In 1988, Harvard University received the first patent for a transgenic animal, a mouse with a human gene for cancer. That animal was called an oncomouse, and it has great usefulness in development of anticarcinogenic drugs. It has been strongly argued that patenting of human

genes is ethically unacceptable. However, the ability to patent genes
and profit from their use is the impetus for funding much of the research. Many companies are seeking to clone medical and agricultur-
ally important genes with the hopes that they will have legal rights
to the use of the information. Is patenting of genes morally acceptable, or is it a just compensation for the work needed to decipher
gene sequence and function? With the differences of opinion, it remains to be seen how this viewpoint will change in society.

HUMAN "RACES" . . . .

Until very recently many people understood that the human population could be subdivided into races according to an individual's skin
color, body shape, hair, and so forth. In 1962, Carleton Stevens Coon,
a famous anthropologist, published the book *The Origin of Races,* sub-
dividing the human population into five races: Caucasian, Mongols,
Australoids, Blacks, and Caboids. However, the biological differences among the people are far more complex than the traits that are
readily seen.

Today, the information from several genetic studies, including the
Human Genome Project, has found that the idea of a human race has
lost much of the biological sense it might have had. According to the
human genome sequencing accomplished by the public consortium
and by the Celera Genomics Corporation, the largest difference between two individuals from any part of the population is about 0.1
percent and the difference between man and a chimpanzee is just
about 1 percent to 2 percent (between a horse and a zebra, the difference at a DNA level is 4 percent). Within chimpanzee species there
are at least three races due to differences among individuals that
evolved in different African forests. However, they look very much
alike despite evolution. In the human species, the genetic difference
among individuals from different continents is much smaller than is
readily apparent. Most of the genetic variability in humans is among
individuals from the same continent, which means that at the DNA
level, some Africans are actually more similar to Europeans or Asians
than they are to other Africans.

Why then, are humans so different in appearance and chimpanzees are not? Chimpanzees originated and remained in tropical

Africa until today. Humans left Africa and dispersed to all continents, occupying the most variable environments on Earth. Humans
continued to adapt to each environment, and the most evident trait
of that adaptation is skin color.

The Human Genome Project has shown that racism does not make
any scientific sense. However, this new biological paradigm relating
to the human race gives the opportunity for other ethical issues to
emerge.

TRADING HUMAN LIFE . . . .

Even before the popularization of the applications of the information
from the Human Genome Project, a changing trend in society can be
seen. Consider, for a moment, a classified ad in *The Minnesota Daily*
published on June 19, 2000 (see Figure 14-2). *The Minnesota Daily* is a

Figure 14-2
University newspaper ad
soliciting an egg donor.
Source: Courtesy of *The Minnesota Daily.*

newspaper at the University of Minnesota, a 150-year-old campus, with more than 64,000 students. The ad offered an $80,000 donation for "an egg to be used in an in vitro fertilization." The couple that was willing to pay that amount certainly had examined other procreation alternatives without success. Obviously, that couple would not accept just any donor. According to the ad, donors were preferred that met the following criteria: height approximately 5'6" or taller, Caucasian, high standardized test scores, college student or graduate under 30, with no genetic medical issues. In compensation, the couple was willing to reimburse the "donation" with a check for the donor or for the charity of her choice. The ad went further, indicating that extra compensation was available for someone who might be especially gifted in athletics, science, mathematics, or music. Additionally, all medical expenses would be covered. This raises some questions: Is there any limit to what money can buy? How much is a human life worth? Seemingly, the words *donation*, *donor*, and *charity* only appear in the text to lessen the ethical implications associated with a monetary reward for a donation.

This is not an isolated case. Similar classifieds were found in other university newspapers, offering up to $100,000 for eggs. Some Web sites offer an online auction of eggs from models, with bids starting at $15,000. These sites have a strong sexual appeal, with sentences inducing customers to succumb to the value of beauty. The site not only shows the donors' pictures, but also has measurements on height, waist, bust, and hips, as well as intelligence, and longevity of parents and grandparents.

Although the harvesting of eggs is a quite safe procedure, a series of risks still exists, from hemorrhage to infertility due to ovarian complications. In spite of risks involved, the "monetary reward" calls into question the validity of this "donation."

Although the idea of donating eggs to help infertile couples is acceptable to a reasonable number of people, the commercialization of this practice is raising ethical issues. Where should the line be drawn for what money can buy? This issue is just the first of many questions appearing due to the progress in reproductive medicine. It is believed that soon egg and sperm donors will also be providing not only information on their physical attributes, but also genomic information highlighting their genetic superiority. Although most in today's society reprove and condemn the trading of eggs and sperm, there are reports of couples buying them at auction.

Societal values might eventually determine which are the good genes and which are the bad genes. The Human Genome Project is a scientific tribute to modern man. However, even considering man a genetic creature, genes by themselves are not enough to create a human. Even if it were possible to put together the genetic puzzle with all of its 30,000 genes, using only the best ones, it would still be impossible to create a human being. A, G, T, and C can define someone's height, skin color, personality, and intelligence, but man is much more than the expression of a sequence of well-organized nucleotide bases in chromosomes. Knowing the complete genome sequence does not lessen the beauty, power, and potential of a human life.

HUMAN CLONING . . . .

Although human cloning has not been done yet, it is believed that it will happen in a matter of time. Since the Dolly sheep cloning in 1997 by Dr. Ian Wilmut at the Roslin Institute in Scotland, the technique has been advanced with many other mammals (monkeys, cows, cats, pigs, etc.). Many countries, however, are passing laws that forbid human cloning. In the United States, California outlawed human cloning in 2001. However, some research groups, mainly in infertility clinics, have indicated their interest in human cloning. Although it is technically possible to clone humans, there are several scientific reasons for not doing so, as discussed in Chapter 6, "Cloning."

Beyond the risks for the pregnant mother and for the clone, a series of ethical issues has been raised in relation to human cloning:

- Would clones have a soul?
- How would clones relate in a family setting or in public settings?
- What would be the limits of paternity and social responsibility to clones?

Consider for a moment the arguments by philosopher George Schedler. Considering the chance of a clone being born with genetic defects and a lawsuit against the parents, Schedler argued, "While society maintains their perception that children with genetic defects are not complete human beings, judges will probably consider them worthy of financial compensation." Some ethicists would argue that cloning violates a child's right to an open future. A cloned child would feel the pressure to become similar to his or her biological donor.

Ethical issues will be raised as society discusses and understands the implications of human cloning. Consider, for example, that human cloning was a reality today. In this scenario a child could have a variable number of parents, from just one to as many as five:

- One parent: This would be the case when a woman has been cloned, serving as the egg donor, the donor of somatic cells, and the surrogate mother.
- Five parents: This would happen when the clone has the following parents:
  - Biological father (somatic cell donor)
  - Biological mother (egg donor)
  - Social father (adoptive)
  - Social mother (adoptive)
  - Surrogate mother

Cloning is a great challenge for society, and moral values certainly will deeply change in the 21st century.

Finally, as humans are not just biological beings, biotechnology should consider its limits on the basis of spiritual values. For example, religious conversions produce profound behavior transformation without any genetic modification. This fact reinforces the idea that human behavior is not just a matter of genes or the environment in which the individual develops. An individual, despite possessing superior genes, can be arrogant and behave irresponsibly in relation to society.

STEM CELLS . . . .

For some people, embryos in their first days of life are nothing more than an amorphous cluster of cells and should not be considered a barrier to research leading to cures for human diseases. For some people, an embryo, beginning at conception, has the status of personhood and deserves social and legal protection. Ethical discussion should precede any conscientious decision. Without ethics, scientific progress in medicine and pharmaceutical development could be underway using individuals instead of animals for early-stage clinical trials, with no regard for human life. If this were the case, the cure for many diseases might have already been discovered, but ethically and legally, human beings cannot be used in scientific experiments that do not have safety standards.

From a biological point of view, the zygote and, consequently, the embryo and fetus are human lives in the most basic form. If they are implanted into a woman, they are able to grow and produce a baby. The ethical question is what the value of a human embryo is. Still, is it possible to balance the interests of millions of people who are suffering from diseases with the interests of an undeveloped embryo? One should not forget that each person, at the beginning of his or her life, developed from an amorphous cluster of cells to be a complete being.

For many, a young mother with small children who was afflicted with Alzheimer's disease would be much more important to save than an embryo, a mass of cells. However, stem cells can be harvested from tissues other than embryos. Scientists prefer the stem cells from embryos because their differentiation potential is higher than those from bone marrow or other tissue. The scientific community and society must proceed with wisdom and prudence. Vanity should not drive unprincipled scientists to dictate the value of human life.

EUGENICS . . . .

Eugenics seeks the genetic improvement of the human race through selection. This idea became more prominent in the beginning of the 20th century and had another great push during World War II. One of Hitler's main objectives was the purification of the Aryan race.

According to the theory of evolution, developed by Charles Darwin in 1859, more "fit" individuals are capable of leaving a larger number of offspring. However, less fit individuals tend to leave fewer descendants. Therefore, over many generations, genes of less adapted or "inferior" individuals are gradually eliminated from the population. Darwin called this process *natural selection*.

Due to the use of medicines and improved medical procedures in the last few centuries, man has evaded natural selection. A classic example is the Cesarean section procedure for childbirth. Women that would otherwise die from natural childbirth can now produce offspring. Therefore, today, genes for body shape that prevent natural deliveries are retained in the population. Natural selection for this trait still exists in indigenous populations that do not have access to modern medicine, but this is the exception and not the rule. In some way, less adapted people with low physical resistance, predisposition for genetic diseases, and so on, continue to leave offspring with the help of modern medical resources. How many of us would not be here if medical resources were not available?

Proponents of eugenics argue that the human species is accumulating bad genes because man has a slow natural selection. Others argue that people need a license for many activities, such as driving, hunting, and fishing, but not to procreate, and therefore the government should also control procreation. In China and India, the government regulates population growth. This is a quantitative and not qualitative control. Some people argue that there is sexual discrimination in those two countries, and that boys tend to be preferred because they can bring larger revenue for their family.

The next 20 years will bring many changes in human behavior, and one can imagine that a revolution could take place that will transform the world. Comparing the world today and that of 50 years ago, no one would think that eugenics could be an issue again. Some of the most despicable human acts were performed in the name of the Aryan race purification. Until 1945, eugenics was taught in many important universities around the world, and the compulsory sterilization of inferior people was relatively common in several countries. There are reports of sterilization of 20,000 people in the United States, 45,000 in England, and 250,000 in Germany during the first half of the 20th century. Eugenics turned public opinion against government intervention in citizens' reproductive choice, and today, compulsory sterilization is only conceivable in the minds of fanatic eugenists. However, with world overpopulation and a growing shortage of resources, many people are afraid that population controls could become a reality again.

Currently, it is difficult to imagine that collective sterilization would be used again, but as genetic tests become routine, it is feared that a new wave of sterilization and abortion could take place, not because of mandatory enforcement, but because of pressure resulting from genetic counseling. Abortion based on genetic counseling is already a reality in countries where it is legal. However, is it right to discriminate against genetic defects or weaknesses, even if it is in the womb? Isn't life just as precious? History has shown that the memory of people is short and that history is cyclical.

With an uncertain future ahead, it seems opportune to recognize that genetic tests and new forms of human reproduction will be part of society from now on. In this scenario, the best alternative seems to be drawing strength from family and moral and religious principles.

BIOTECHNOLOGY AND CHRISTIAN FAITH . . . . . .

The Bible says that man was created in the image of God. What is said about the image of God in the Bible? Passages in the Book of Genesis and one in the New Testament address this topic. In Genesis 1:26-27, "I will make man in My image, after My likeness . . . And God created man in His image, in the image of God He created him; male and female created He them," image is introduced in a parallel to man's likeness to God. Just being the image and likeness of God is enough to make people recognize that their life is sacred (Genesis 9:6). In the New Testament, Paul emphasizes the moral dimension of creation and the image in which man was made. The new man in Christ is a transformed image of its Creator (Ephesians 4:24).

Although the Bible does not discuss man's biology, it is clear that human beings should be seen with God's perspective. Many of the ethical principles of society reflect its religious beliefs. Most Western countries have established ethical principles based in Christianity. This can help many to form opinions on applications of biotechnology for human beings, but does not address many of the other uses for this science.

Human Genome and Religious Considerations

The genome sequencing of many living organisms—the determination of sequence of the genetic alphabet (A, C, G, and T)—is transforming genetics from science to an area of great economic activity. The budget for the human genome-sequencing project done by the public sector is estimated at $3 billion. About 3 percent of this budget was allocated to studying the ethical implications of the sequencing. The sequencing carried out by the private sector (Celera Genomics) is also using a significant volume of its resources to address ethical issues. Even with a group of scholars in ethics backing up all the information from the project, many questions remain unanswered. The new knowledge about the human genome is shaping a new perception of many old concepts like genes, genetic destiny, and human nature. Many specialists are referring to genome sequence as the "Book of Life." Some of the analogies to this book with its 3.2 billion letters indicate that it would take a century for one to recite it at the speed of one letter per second, 24 hours a day, 365 days a year.

With the first draft of the human genome almost completed, the anticipation that the mysteries of life are to be solved is huge. This perspective carries the idea that genome and human nature are synonymous. According to this perception, the genome is understood as being the human essence, and environment is a factor that modifies the traits coded for by each genome. Assuming that the genome is a complete receipt for a human being is a great mistake, and it neglects all metaphysical aspects and the spirituality of humans. This perspective excludes the spirit as an inherent part of a human being. This revolutionary idea reverts to the theme of Aristotle and other philosophers from the Middle Ages. Thomas Aquinas considered the beginning of organization in living organisms as a concept of form, which shapes its traits. The moment that man assumes this form is seen in different ways. Whereas some consider conception to be the beginning of life, others think that an amorphous conglomerate of cells (zygote and embryo) does not deserve the status of personhood.

Part of the rationale that the genome is synonymous with human nature appears from ideas on what human nature and genes are. If genes code for an individual's traits (e.g., eye and skin color, voice type), the importance of the genome in determining what constitutes an individual is evident. The discovery of genes for alcoholism, violence, and homosexuality seems to reinforce that human nature is defined by the genome. Realizing that within the genome, genes are present that confer musical gifts, mannerisms, and personality, it seems that there is no space left for other considerations.

One of the conceptual mistakes of metaphysical genomics is the issue of when a human being becomes a person, with all of its human rights. Most people acknowledge that at birth, a child already has basic rights and that a spermatozoid does not have them. Therefore, the issue is when, in the development of a baby, he or she acquires the status of a person, which can be understood as the human essence. For many, this moment coincides with the fertilization of the egg, which possesses 23 chromosomes, by the spermatozoid, which also possesses 23 chromosomes. This fertilization or conception results in a zygote, the first stage of the human life, in which the number of chromosomes is recovered ($2n = 46$). Conception is the moment at which the human genome becomes complete and is able to start the development of a new individual. This also seems to support the idea that the genome is the same as human essence.

For those that agree that the zygote is the beginning of human life, the notion developed by Thomas Aquinas becomes an anachronism, because it suggests that a human embryo only receives its human essence when it is sufficiently formed. Some argue that this would happen on the 14th day after conception, when formation of the brain begins. Others argue that this would only occur at the 40th day for boys and at the 90th day for girls, based on the development of embryo. Finally, some believe that this only happens at birth.

Some of the most recent perspectives on form include the idea that form is not appearance but it defines essence. The human form confers structure and human material as soon as the genetic information (in-*form*-ation) necessary to define an individual is put together. The way the word *information* can be dissected reveals how the genome can be viewed. Thomas Aquinas was, therefore, mistaken about the exact moment that the human essence is attributed to man. At the time, Thomas Aquinas did not have the scientific knowledge available in the 21st century.

In this discussion, an aspect that needs to be addressed is that personal identity does always not agree with genomic identity. For example, identical twins come from a single zygote and they possess exactly the same genome, but they are different people. Despite having the same physical and psychological characteristics and the same predisposition for genetic disease, identical twins are separate people. Although experts in the area of cloning argue that the same genome does not mean the same person, many still do not seem to understand this idea.

Biotechnology has brought about many previously unimaginable possibilities that society needs to address. For instance, the sequencing of the human genome revealed that the difference between the human genome and that of a chimpanzee is only 1.3 percent. This finding left scientists, philosophers, and the public perplexed. It is not because chimpanzees are 98.7 percent similar to humans, at the genomic level, that they are inferior primates. Obviously that small genomic difference is enough to set *Homo sapiens* apart as a peculiar species. Although genes are essential for human existence, they are not enough.

Attributing all our hopes and fears to genes is an elementary mistake. Biotechnology allows greater control of human destiny, but this does not mean that to be called human it is enough to possess 46 chromosomes with their 3.2 billion nucleotides, for science alone cannot explain life. The complexity written from only four letters of the genetic alphabet is just another piece of evidence for the hand of a Creator.

Experience teaches that the more that is known, the more is to be learned. Finally, it can be expected that the 21st century will bring additional evidence of the complexity and the simpleness of creation. No matter how much biotechnology reveals about man's biological essence, the inclusion of a spiritual dimension will always be needed for the picture to be complete and for ethics to remain focused.

CASE STUDIES . . . .

If something is harmful, we tend to avoid it. This simple rule might work well in theory, but not in practice. What is the best way to behave in situations in which the usefulness of something is not clear or when the decision could be both useful and harmful? Biotechnology can create situations in which such ambiguity prevails. Most of the differences in opinion between those in favor of biotechnology and those opposed to it are due to differences in their perceptions and interpretations of the risks and benefits. The two following hypothetical cases show that biotechnology is making information available that could bring uncertainty and discovery of new issues to many people.

Case 1

Genetic tests for many diseases have existed for more than 20 years, and they have been used in family planning.

Junior is a 6-year-old boy, the son of Paul and Anne. He was diagnosed with cystic fibrosis at the age of 2. Cystic fibrosis is a genetic disease that occurs in individuals that carry two recessive alleles (*aa*), one coming from each parent. Individuals with the *AA* or *Aa* alleles do not have the disease.

Anne spends a great part of her time taking care of Junior, assuring the expectoration of accumulated mucus in his lungs. Additionally, Junior takes 15 pills a day as part of his therapy. On average, Junior is hospitalized twice a year to fight breathing infections. With daily care, Junior might live to be 30 years old, but had he been born 15 years ago, he would have likely died early.

Paul and Anne are planning to have a second child. They are advised by Junior's doctor to take a genetic test for cystic fibrosis before Anne becomes pregnant. The results indicate that Anne is a carrier of the gene, but Paul is not. As Paul is not a carrier, it is impossible that he is Junior's biological father.

Ethical Dilemmas

1. Was it morally justified for the doctor to recommend a genetic test?
2. Should the doctor present the test result to Paul and/or Anne?
3. Should the doctor decide not to show the results, but Paul and Anne insist, should the doctor speak the truth or hide it?

Possible Solutions

Considering that the doctor has decided to let the parents know the results of test, there would be many alternatives:

1. Report the results to Paul and let him decide what to do.
2. Report the results to Anne and let her decide what to do.
3. Report only what is asked.
4. Report the results to Junior when he turns 18.
5. Report the results to both parents and let them work it out.
6. Other alternatives.

None of these options is perfect. If the doctor shares the good news that the couple can have children without disease, that would raise questions about Junior's paternity. This is typically an ambiguous situation because benefits and risks exist associated with the revelation of the genetic test results. The benefit of revealing the results is that Anne bears another child with no risks of cystic fibrosis. The risk would be the destruction of their family, harming each member.

Although the doctor's option to reveal the results just to Anne and let her have the option to share them with Paul is not perfect, at least the rights of both parents would be respected. In that case, the rationale that guided the doctor's decision was to preserve the family.

Case 2

Sonja is a healthy and gifted accountant. She decides to work for a large, respected accounting firm.

Five years ago, Sonja's mother died of Huntington's disease, which is caused by a dominant mutant gene, meaning only one copy of the gene is needed to cause the disorder. This is a neurodegenerative disease that becomes evident during middle age. The disease results in insanity and dementia and causes early death in those affected. There is no cure for this disease; however, genetic tests can identify those carrying the gene. The knowledge that an individual is a carrier of the mutated gene can be the same as a death sentence. The onset of the disease usually occurs in middle age, around age 45 to 50, as in Sonja's mother's case. The diagnosis of Sonja's mother's disease was restricted to her doctor and family. After genetic and psychological counseling, none of Sonja's siblings decided to take the genetic test.

Sonja was selected to work for the company of her dreams. During the completion of the recruiting forms, she discovered that she would have to submit to clinical examinations and genetic tests.

When she questioned the need to submit herself to the DNA tests, she was informed that the company required the procedure.

Ethical Dilemmas

1. Is it ethically moral for the company to demand genetic tests?
2. Who should have access to Sonja's results?
3. Could the company refuse employment to Sonja on the basis of a positive result?
4. If the company where Sonja's brother works has access to her genetic test, can it also be used to discriminate against him?

Consider the first dilemma, which gives us the opportunity to discuss legal and ethical behavior issues. Even if the law does not prohibit companies from demanding genetic tests during employee recruiting, it does not mean that requesting such tests is ethically acceptable.

Possible Solutions

1. The company does not have the ethical right to demand the tests.
2. The company should only demand the tests if it is trying to prevent possible harm to its employees, such as preventing susceptible individuals to chemical exposure.
3. The company could request the test just for insurance purposes, but not for recruiting issues.

4. Sonja should seek employment in another company.
5. Sonja should take the test in an independent laboratory and certify that only she will have access to the results.

In this case, even if the test result is positive, the onset of the disease could occur after retirement, but it would likely have a psychological impact on the individual with a reflection on his or her job performance and also psychologically affect other siblings.

FINAL CONSIDERATIONS . . . .

The fine line between right and wrong, or between ethically acceptable and ethically unacceptable behavior, is a tremendous part of bioethics. If it were possible to define those limits or present a rule of thumb to guide ethically correct decisions, it would certainly be mentioned here. However, it seems that many of the ethical cases related to biotechnology have no clear right or wrong and should be judged on an individual basis. Ethical considerations relating to the many facets of biotechnology should not be something that is only discussed at a company board meeting or within the committee rooms of governing organizations. Many of these new technologies will affect everyone in the near future, and it is important to recognize the many considerations involved in biological sciences. Biotechnology is advancing and making progress on the major factors that limit the lives of billions of people. Solutions to the problems of hunger, disease, pollution, and others are being found using the science of biotechnology, yet many are apprehensive about the technology or fear the technical nature of the science.

Despite the greatest efforts, balanced arguments that will satisfy everyone are impossible to find. The first step for anyone is to become educated in the background, the science, and the applications of biotechnology. It is then important to be informed about how biotechnology affects the risks, benefits, and moral implications associated with superior health care, enhanced crop production, and environmental improvement. This knowledge must be used to make informed and sound judgments, so that opinions are based on fact and study, and not on emotion, hype, or fear. Ultimately, each individual must take the essential steps to understand biotechnology.

Glossary

Agrobacterium rhizogenes
Soil bacteria that often possesses the Ri plasmid, and might cause the hairy root symptom in plants. It is used in plant transformation.

Agrobacterium tumefaciens
Soil bacteria that often possesses the Ti plasmid, and might cause the crown gall tumor in plants. It is used in plant transformation.

Allele
One of a number of different forms of a gene. Each person inherits two alleles for each gene, one from each parent. These alleles might be the same or might be different from one another.

Amino acid
One of 20 different molecules that combine to form proteins. The sequence of amino acids in a protein determines the protein structure and function.

Anticodon
The triplet of nucleotides in transfer RNA, which associates by complementary base pairing with a specific triplet (codon) in messenger RNA during protein synthesis.

Antisense gene
Gene in the reverse orientation in relation to the promoter, which when transcribed, produces a complementary polynucleotide to the gene in the original orientation.

Apoptosis
The process that, when functioning normally, programs cells to self-destruct at an appropriate moment in an organism's life cycle. If the apoptotic process malfunctions in a cell, uncontrolled cell growth can result, which can contribute to the development of cancer. Such disruption of apoptosis might be associated with an inherited genetic mutation or a somatic cell genetic mutation.

BAC (bacterial artificial chromosome)
A vector used to clone DNA fragments of 100 to 300 kb insert size (average of 150 kb) in *Escherichia coli* cells. Based on the naturally occurring F-factor plasmid found in the bacterium *Escherichia coli.*

Bacillus thuringiensis (Bt)
An insecticidal soil bacterium, marketed worldwide for control of many important plant pests, mainly caterpillars of the genus *Lepidoptera* (butterflies and moths), but also for control of mosquito larvae. The toxin genes from Bt have also been genetically engineered into several crop plants.

Base
One of the molecules (adenine, guanine, cytosine, thymine, or uracil) that form part of the structure of DNA and RNA molecules. The order of bases in a DNA molecule determines the structure of proteins encoded by that DNA. See also *nucleotide*.

Base pair (bp)
Two complementary nucleotide bases joined together by chemical bonds. The two strands of the DNA molecule are held together in the shape of a double helix by the bonds between base pairs. The base adenine pairs with thymine, and guanine pairs with cytosine.

Biodegradation
The process whereby a compound is decomposed by natural biological activity. A biodegradable substance is one in which a microbe can reduce it to CO_2 and H_2O by mineralization.

Biodiversity
All living organisms in an ecosystem.

Bioinformatics
The science of informatics as applied to biological research. Informatics is the management and analysis of data using advanced computing techniques. Bioinformatics is particularly important as an adjunct to genomics research because of the large amount of complex data this research generates.

Biolistic
A method (biological ballistics) of transforming cells by bombarding them with microprojectiles coated with DNA.

Bioreactor
Individuals used for production of biotech products such as proteins, enzymes, and so on.

Bioremediation
The use of biological agents to render hazardous wastes nonhazardous or less hazardous.

Biosafety
Science that deals with the evaluation, control, and minimization of risks from biotechnology. Biosafety is concerned with risks to human and animal health and to the environment.

Biotechnology
The application of biological research techniques to the development of products or processes, using biological systems, living organisms, or derivatives thereof. It can also be understood as a range of different molecular technologies such as gene

manipulation and gene transfer, DNA typing, and cloning of microorganisms, plants, and animals.

bp (base pair)
See *base pair (bp)*.

BRCA (breast cancer gene)
Genes that normally help to restrain cell growth, but can contain certain genetic mutations associated with the development of breast and ovarian cancer. Note, however, that inherited BRCA1 and BRCA2 mutations are thought to account for less than 10 percent of all breast and ovarian cancers. Recent evidence suggests that somatic cell genetic mutations (i.e., noninherited genetic mutations) in these two genes could also play a role in the development of cancer.

BST (bovine somatropine)
Bovine growth hormone. BST is similar to human growth hormone and it is used for improving milk production in cows.

Bt
Bacillus thuringiensis. Bt varieties, developed through biotechnology, are carriers of genes for *cry* or Bt protein, which are lethal to many insects like the European corn borer and others, but are not harmful to mammals.

cDNA (complementary DNA)
DNA that is synthesized in the lab from mRNA templates.

Center of diversity
Geographic region that has the highest genetic diversity of a species. Sometimes also called *center of origin*.

Central dogma
The original postulate that genetic information can be transferred only from nucleic acid to nucleic acid and from nucleic acid to protein; that is, from DNA to DNA, from DNA to RNA, and from RNA to protein (although information transfer from RNA to DNA was not excluded and is now known to occur; this is reverse transcription). Transfer of genetic information from protein to nucleic acid never occurs.

Chromomere
One of the serially aligned beads or granules of a eukaryotic chromosome, resulting from local coiling of a continuous DNA thread.

Clone
A term applied to genes, cells, or entire organisms that are derived from, and are genetically identical to, a single common ancestor gene, cell, or organism, respectively. Cloning of genes and cells to create many copies in the laboratory is a common procedure essential for biomedical research. Several processes, which are commonly described as cell cloning, give rise to cells, which are almost but not completely genetically identical to the ancestor cell. Cloning of organisms from embryonic cells occurs naturally in nature (e.g., with the occurrence of identical twins). The laboratory cloning of a sheep using the genetic material from a cell of an adult animal has recently been reported. This is the production of an animal with nuclear DNA identical to another animal. Cloning takes place without the contribution of DNA from a male and

female, and is therefore "asexual" reproduction.

Cloning
The process of producing a genetically identical copy.

Codon
A set of three nucleotide bases in a DNA or RNA sequence, which together code for a unique amino acid. For example, the set AUG (adenine, uracil, guanine) codes for the amino acid methionine.

Construct
Plasmid having a promoter, the coding sequence, and a terminator, that is introduced in a host via biolistic, *Agrobacterium*, microinjection, and so on, to transform it.

Contigs
Groups of cloned DNA sequences representing overlapping regions of a genome.

Cosmid
An artificially constructed cloning vector/plasmid that contains cos sites at each end. Cos sites are recognized during head filling of lambda phages. Cosmids are useful for cloning large segments of foreign DNA.

cRNA (complementary RNA)
Synthetic RNA produced by transcription from a specific DNA single-stranded template.

Differentiation
The process by which a primitive cell commits to becoming a specialized cell in the body, such as a skin cell or a bone cell.

DNA (deoxyribonucleic acid)
The molecule that encodes genetic information. DNA is a double-stranded helix held together by bonds between pairs of nucleotides.

DNA ligase
An enzyme that closes nicks or discontinuities in one strand of double-stranded DNA by creating an ester bond between adjacent 3' OH and 5' PO_4 ends on the same strand.

DNA polymerase
An enzyme that can synthesize new DNA strands using a DNA template; several such enzymes exist. There are several classes of enzymes that polymerize DNA nucleotides using single- or double-stranded DNA as a template.

DNA vaccine
Product with naked DNA used to induce an immune response against the pathogen from which the DNA was obtained.

Dominant
An allele that determines phenotype even when heterozygous. Also the trait controlled by that allele.

Electroporation
Technique that uses an electric discharge to produce pores on the cell membrane for intake of recombinant DNA.

Environment
The combination of all the conditions external to the genome that potentially affect its expression and its structure.

Enzyme
A protein that facilitates a biochemical reaction in a cell. In general, these biochemical reactions would not occur if the enzyme were not

present. For example, an enzyme can facilitate (also called *catalyze*) the destruction of another protein by breaking the bonds between amino acids. An enzyme of that type is called a *protease*.

Escherichia coli
A common bacterium that has been studied intensively by geneticists because of its small genome size, normal lack of pathogenicity, and ease of growth in the laboratory.

Eucaryote
Multicellular and unicellular organisms (yeast, plants, animals, etc.), with cells that have a distinct nucleus, enveloped by a nuclear membrane.

Eugenics
Controlled human breeding based on notions of desirable and undesirable genotypes.

Evolution
In Darwinian terms, a gradual change in phenotypic frequencies in a population that results in individuals with improved reproductive success.

Gene
A length of DNA, which codes for a particular protein or, in certain cases, a functional or structural RNA molecule.

Gene bank
A group of genes or cloned DNA fragments, which are coordinately controlled.

Gene cloning
Isolating a gene and making many copies of it, which is accomplished by inserting the gene's DNA se-

quence into a vector, transmitting the transgenic construct into a cell, and then allowing the cell to reproduce, thereby creating many identical copies of the gene.

Gene escape
See *gene flow*.

Gene flow
The exchange of genetic information among individuals, populations, or species, with the preservation of that genetic information in the following generations.

Gene myth
Idea that every trait is determined by genes (i.e., stature, eye color, tendency to crime, etc.).

Gene reporter
Gene usually inserted in the construct to facilitate the visual identification of transformed tissue. Examples are genes for glucaronidase (GUS), luciferase, or green fluorescent protein (GFP).

Gene silencing
Partial or total suppression of gene expression.

Gene therapy
The correction of a genetic deficiency in a cell, tissue, or organ by the addition of a normal, functioning copy of a gene and its insertion into the genome. This technique should be clearly distinguished from the use of genomics to discover new targets for drug discovery and new diagnostic tools.

Genetic code
The set of codons in DNA or mRNA. Each codon is made up of three nu-

cleotides, which call for a unique amino acid. For example, the set AUG (adenine, uracil, guanine) calls for the amino acid methionine. The sequence of codons along an mRNA molecule specifies the sequence of amino acids in a particular protein.

Genetic engineering
Altering the genetic material of cells or organisms to make them capable of making new substances or performing new functions.

Genetic erosion
Loss of genetic diversity and variability caused by either natural or man-made processes.

Genetic map
A map of a genome that shows the relative positions of the genes and/or markers on the chromosomes.

Genetic pollution
Uncontrolled escape of genetic information (frequently referring to products of genetic engineering) into the genomes of organisms in the environment where those genes never existed before.

Genetic vulnerability
Condition of low genetic diversity that might predispose an organism to susceptibility to pests.

Genome
All the genetic material in the chromosomes of a particular organism; its size is generally given as its total number of base pairs.

Genomics
The science of identifying the sequence of DNA in various species, and subsequent processing of that information, or the study of all genes and their function. Recent advances in genomics are bringing about a revolution in our understanding of the molecular mechanisms of disease, including the complex interplay of genetic and environmental factors. Genomics is also stimulating the discovery of breakthrough health products by revealing thousands of new biological targets for the development of drugs and by giving scientists innovative ways to design new drugs, vaccines, and DNA diagnostics. Genomics-based therapeutics include "traditional" small chemical drugs, protein drugs, and potentially gene therapy.

Germ cell
Sperm and egg cells, and their precursors. Germ cells are haploid and have only one set of chromosomes (23 in humans), whereas all other cells have two copies (46 in humans).

Germplasm
All the genetic background of a species.

GMO
Genetically modified organism. Loosely defined, it is an organism that has been changed from its natural state. Some define a GMO as an organism in which foreign DNA has been inserted using recombinant DNA technology.

Golden Rice
Genetically modified rice variety developed by Ingo Potrykus and Peter Beyer at the Swiss Federal Institute

of Technology and University of Freiburg, Germany. Golden Rice has a high content of β-carotene (a vitamin A precursor).

GUS (β-glucaronidase)
Gene from *Escherichia coli* used as reporter in transformation.

Herbicide
Product used to kill plants.

Homologue
This term is used by geneticists in two different senses: one member of a chromosome pair in diploid organisms, or a gene from one species (e.g., the mouse) that has a common origin and functions the same as a gene from another species (e.g., humans, *Drosophila*, or yeast).

In vitro
Pertaining to a biochemical process or reaction taking place in a test tube (or more broadly, in a lab) as opposed to taking place in a living cell or organism.

In vivo
Pertaining to a biological process or reaction taking place in a living cell or organism.

Insert
Exotic DNA fragment introduced into a vector.

Interferon
Protein naturally produced by human body cells that increases the resistance to virus infection. Alpha-interferon is known to be effective against some cancers.

kb
A length of DNA equal to 1,000 nucleotides.

Marker
A sequence of bases at a unique physical location in the genome, which varies sufficiently between individuals that its pattern of inheritance can be tracked through families and it can be used to distinguish among cell types. A marker might or might not be part of a gene. Markers are essential for use in linkage studies and genetic maps to help scientists narrow down the possible location of new genes and discover the associations between genetic mutations and disease.

Microarray
Glass slide with a large number of DNA fragments used for probing biological samples for different attributes and effects.

Mutation
A change, deletion, or rearrangement in the DNA sequence that could lead to the synthesis of an altered inactive protein or the loss of the ability to produce the protein. If a mutation occurs in a germ cell, then it is a heritable change and can be transmitted from generation to generation. Mutations can also be in somatic cells and are not heritable in the traditional sense of the word, but are transmitted to all daughter cells.

Nuclear transfer
A general term for the process of cloning where the genetic information from a somatic cell is transferred to an egg cell from which the DNA is removed.

Nucleic acid
One of the family of molecules that includes the DNA and RNA

molecules. Nucleic acids were so named because they were originally discovered within the nucleus of cells, but they have since been found to exist outside the nucleus as well.

Nucleotide

The building block of nucleic acids, such as the DNA molecule. A nucleotide consists of one of four bases (adenine, guanine, cytosine, or thymine) attached to a phosphate-sugar group. In DNA the sugar group is deoxyribose, whereas in RNA (a DNA-related molecule that helps to translate genetic information into proteins), the sugar group is ribose, and the base uracil substitutes for thymine. Each group of three nucleotides in a gene is known as a *codon*. A nucleic acid is a long chain of nucleotides joined together, and therefore is sometimes referred to as a *polynucleotide*.

Oncogene

A gene that is associated with the development of cancer.

Patent

A grant issued by the government that gives the patent holder the right to exclude others from making, using, or selling a patented invention for a certain term. In most countries, including the United States, the term begins on the date on which the patent is issued and ends 20 years from the date on which the application for the patent was filed. In the United States, patents are granted on inventions that meet the requirements of novelty, nonobviousness, and utility. A patent holder cannot use a patented invention dominated by the patent of another, absent a license or cross-license.

PCR (polymerase chain reaction)

A powerful method for amplifying specific DNA segments that exploits certain features of DNA replication. For instance, replication requires a primer and specificity is determined by the sequence and size of the primer. The method amplifies specific DNA segments by cycles of template denaturation, primer addition, primer annealing, and replication using a thermostable DNA polymerase. The degree of amplification achieved is set at a theoretical maximum of 2^n, where n is the number of cycles (e.g., 20 cycles gives a theoretical 1,048,576-fold amplification).

Pharmacogenomics

Science that studies the correlation between the genome of an individual and his or her response to drugs used in treatments.

Plasmid

Autonomously replicating, extra-chromosomal circular DNA molecules, distinct from the normal bacterial genome and nonessential for cell survival under nonselective conditions. Some plasmids are capable of integrating into the host genome and are used as a cloning vector for small pieces of DNA.

Primer

Short pre-existing polynucleotide chain to which DNA polymerase can add new deoxyribonucleotides.

Prion (proteinaceous infected particle)

Proteinaceous molecules found in the cerebral cell membrane of vertebrate animals. Mutated forms of those molecules might cause neural degenerative diseases, such as mad cow disease.

Promoter

A segment of DNA located at the "front" end of a gene, which provides a site where the enzymes involved in the transcription process can bind to a DNA molecule and initiate transcription. Promoters are critically involved in the regulation of gene expression.

Protein

A biological molecule that consists of many amino acids linked together by peptide bonds. The sequence of amino acids in a protein is determined by the sequence of nucleotides in a DNA molecule. As the chain of amino acids is being synthesized, it is also folded into higher order structure shapes (e.g., helices or flat sheets). Proteins are required for the structure, function, and regulation of cells, tissues, and organs in the body.

Proteomics

The science of identifying the sequence, function, structure, and interrelationship of all proteins in an organism and subsequent processing of that information.

Recombinant DNA

DNA molecules that have been created by combining DNA from more than one source.

Regenerative medicine

The creation and transplantation of healthy cells, tissues, and organs to replace or repair a patient's own damaged or diseased cells, tissues, and organs.

Restriction enzyme

A protein that recognizes specific, short nucleotide sequences and cuts DNA at those sites. Bacteria contain more than 400 such enzymes that recognize and cut over 100 different DNA sequences.

RNA (ribonucleic acid)

A molecule similar to DNA, which helps in the process of decoding the genetic information carried by DNA.

Sequencing

Determining the order of nucleotides in a DNA or RNA molecule or determining the order of amino acids in a protein.

Somatic cell

Any cell in the body except gametes and their precursors. In diploid individuals, somatic cells possess $2n$ chromosomes.

Stem cell

Undifferentiated, primitive cells in the bone marrow with the ability to both multiply and differentiate into specific blood cells.

Terminator

The site on a DNA sequence at which transcription stops.

Tissue culture

Term used for in vitro growth of cells, tissue, or organs, in aseptic conditions and in a nutritive media.

Totipotent

Cells that have the ability to develop into any of the many different cell types that make up multicellular organisms. Embryos are composed of large numbers of totipotent cells, which decline in number as development proceeds and cell specialization begins to occur. Adults have a much more limited ability to produce totipotent cells than do embryos. Organisms such as humans retain a complete set of genetic information in all adult body cells, but only a small fraction of an adult's cells have the ability to develop into multiple cell types. Recent research has shown that differentiated adult cells can be treated such that they become totipotent. Such totipotent or stem cells offer possibilities for a number of therapeutic uses, such as repairing heart muscle after a heart attack or brain function after a stroke. Plant cells tend to retain a greater capability for totipotence, even in mature plants, than do those of animals.

Transcription

The process during which the information in a sequence of DNA is used to construct an mRNA molecule.

Transformation

A process by which the genetic material carried by an individual cell is altered by incorporation of exogenous DNA into its genome.

Transgenic

An organism with a genome that has been altered by the inclusion of foreign genetic material. This foreign genetic material can be derived from other individuals of the same species or from wholly different species. Genetic material can also be artificial. Foreign genetic information can be added to the organism during its early development and incorporated in cells of the entire organism. For example, mice embryos given the gene for rat growth hormone develop into larger adults. Genetic information can also be added later in development to selected portions of the organism. As an example, experimental genetic therapy to treat cystic fibrosis involves selective addition of genes responsible for lung function and is administered directly to the lung tissue of children and adults. Transgenic organisms have been produced that provide enhanced agricultural and pharmaceutical products. Insect-resistant crops and cows that produce human hormones in their milk are but two examples.

Translation

The process during which the information in mRNA molecules is used to construct proteins.

Vector

An organism that serves to transfer a disease-causing organism (pathogen) from one organism to another. It is also a mechanism whereby a foreign gene is moved into an organism and inserted into that organism's genome. Retroviruses such as HIV serve as vectors by inserting genetic information (DNA) into the genome of human cells. Bacteria can serve as vectors in plant populations.

Virus

A piece of nucleic acid covered by protein. Viruses can only reproduce by infecting a cell and using the host's cellular mechanisms for self-replication, and can cause disease. Modified viruses can also be used as a tool in gene therapy to introduce new DNA into a cell's genome.

Xenotransplant

Transplantation of tissue or organs between organisms of different species, genus, or family. A common example is the use of pig heart valves in humans.

YAC (yeast artificial chromosome)

A vector used to clone DNA fragments (up to 400 kb); it is constructed from the telomeric, centromeric, and replication origin sequences needed for replication in yeast cells. Compare *cosmid*.

Yeast

Fungus from the *Saccharomycetaceae* family, which is used in fermentation of liquor and in bread making.

Zygote

A cell produced by the fusion of a female gamete (egg cell or ovum) with a male gamete (sperm cell or pollen grain). The joining of a sperm and egg cell is called *fertilization*. Zygotes are diploid and undergo cell division to become an embryo.

References

Alcamo, E. (1999). *DNA technology: The awesome skill.* New York: Harcourt Academic Press.

Andrade, A. (1989, March). A Tutela ao meio ambiente e a constituição (Tutoring the environment and the constitution). *AJURIS, 45.*

Ballantyne, J., Sensabaugh, G., & Witkowski, J. (1989). *DNA technology and forensic science.* New York: Cold Spring Harbor Laboratory.

Borém, A. (2001). *Escape gênico e transgênicos (Gene escape and transgenics).* Rio Branco, Brazil: Editora Suprema.

Borém, A. (2001). *Melhoramento de plantas (Plant breeding)* (3rd ed.). Viçosa, Brazil: Editora UFV.

Calixto, J. B. (2001). Biopirataria (Biopiracy). *Ciência Hoje, 28,* 36–43.

Carvalho, A. C. C. (2001). Células tronco: A medicina do futuro (Stem cells and the future of medicine). *Ciência Hoje, 29,* 26–31.

Daniell, H., Streatfield, S. J., & Wycoff, K. (2001). Medical molecular farming: Production of antibodies, biopharmaceuticals and edible vaccines in plants. *Trends in Plant Science, 6,* 219–226.

Drlica, K. (1996). *Understanding DNA and gene cloning: A guide for the curious* (3rd ed.). New York: Wiley.

Giddings, G., Allison, G., Brooks, D., & Carter, A. (2000). Transgenic plants as factories for biopharmaceuticals. *Nature Biotechnology, 18,* 1151–1155.

James, C. (2001). *Global status of commercialized transgenic crops: 2000.* Available at *http://www.isaaa.org/publications/briefs/Brief_21.htm*

Koprowski, H., & Yusibov, V. (2001). The green revolution: Plants as heterologous expression vectors. *Vaccine, 19,* 2735–2741.

Lackie, J. M., & Dow, J. (2000). *The dictionary of cell and molecular biology*. New York: Academic.

Langridge, W. H. R. (2000). Edible vaccines. *Scientific American, 283*, 66–71.

Leite, M. (2000). *Os alimentos trangênicos (Genetically modified foods)*. São Paulo, Brazil: Publifolha.

Lewin, B. (1999). *Genes VII*. Oxford, England: Oxford University Press.

Mason, J. H., & Arntzen, C. J. (1995). Transgenic plants as vaccine production systems. *Trends in Biotechnology, 13*, 388–392.

Mercenier, A., Widermann, U., & Breiteneder, H. (2001). Edible genetically modified microorganisms and plants for improved health. *Current Opinions in Biotechnology, 12*, 510–515.

Messina, L. (2000). *Biotechnology*. New York: Wilson.

Perelman, C. (1999). *Ética e direito (Ethics and law)*. São Paulo, Brazil: Editora Martins Fontes.

Phillips, R. L., & Vasil, I. K. (2001). *DNA-based markers in plants* (2nd ed.). New York: Kluwer Academic.

Quist, D., & Chapela, I. (2001). Transgenic DNA introgressed into traditional maize landraces in Oaxaca, Mexico. *Nature, 414*, 541–543.

Rae, S., & Cox, P. (2001). *Bioethics: A Christian approach in a pluralistic age*. New York: Kluwer.

Rocha, M. M. (1998). *Biotecnologia e patentes (Biotechnology and patents)*. In A. Borém et al. (Eds.), *Biowork*. Viçosa, Brazil: Imprensa UFV.

Snustad, P., & Simmons, M. J. (1999). *Principles of genetics* (2nd ed.). New York: Wiley.

Tabashnik, B. E. (1994). Reversal of resistance to *Bacillus thuringiensis* in *Plutella xylostella*. *Proceedings of the National Academy of Science USA, 91*, 4120–4124.

Tang, J. D. (1996). Toxicity of *Bacillus thuringiensis* spore and crystal to resistant diamondback moth (*Plutella xylostella*). *Applied Environmental Microbiology, 62*, 564–569.

Tourinho Neto, F. (1997). *A constituição na visão dos tribunais (The vision of the constitution in the courts of law)*, vol. 3. Tribunal Regional Federal – 1ª Região.

Ueta, J., Pereira, N. L., Shuhama, I. K., & Cerdeira, A. L. (2000). *Biodegradação de herbicidas e biorremediação: Microrganismos degradadores do herbicida atrazina (Herbicide biodegradation and bioremediation: Degrading microorganisms for the herbicide atrazin)*. http://www.biotecnologia.com.br/bio/10_i.htm.

Varella, M. D., Fontes, E., & Rocha, F. G. (1999). *Biossegurança e biodiversidade—contexto científico e regulamentar (Biosafety and biodiversity—scientific and regulatory aspects)*. Belo Horizonte: Editora Del Rey.

Watson, J. D. (2000). *A passion for DNA: Genes, genomes and society.* New York: Cold Spring Harbor Laboratory Press.

Watson, J. D., Gilman, M., & Witkowski, J. (1992). *Recombinant DNA* (2nd ed.). New York: Freeman.

Wilmut, I., Campbell, K. E., & Tudge, C. (2000). *The second creation: Dolly and the age of biological control.* Cambridge, MA: Harvard University Press.

SUGGESTED WEB SITES . . . .

ABS Global: *http://www.absglobal.com/*

Access Excellence: *http://www.accessexcellence.org/AB/*

Ag Biotech Infonet: *http://www.biotech-info.net/*

Articles on DNA: *http://www.dnafiles.org/home.html*

Biodiversity Information Network, Brazil: *http://www.binbr.org.br*

Bioethics Center: *http://www.bioethics.umn.edu/*

Bioethics Net: *http://www.med.upenn.edu/bioethic/*

Biota: *http://www.biota.org.br/*

Celera Genomics: *http://www.celera.com*

Cold Spring Harbor Laboratory: *http://vector.cshl.org/*

Convention on Biological Diversity: *http://www.biodiv.org/*

Council for Biotechnology Information: *http://www.whybiotech.com/*

CTNBio: *http://www.ctnbio.gov.br/ctnbio/default.htm*

Dictionary of Life Science: *http://biotech.icmb.utexas.edu/search/dict-search.html*

GenBank (NCBI): *http://www.ncbi.nlm.nih.gov*

Georgia Bureau of Investigation: *http://www.ganet.org/gbi/fsdna.html*

Greenpeace: *http://www.greenpeace.org/~geneng/*

Human Genome Project: *http://www.ornl.gov/TechResources/ Human_Genome/home.html*

Information System for Biotechnology: *http://www.nbiap.vt.edu/*

International Service for the Acquisition of Agri-biotech Applications: *http:// www.isaaa.org*

National Center for Biotechnology Education: *http://www.ncbe.reading.ac.uk/*

Plant Biotechnology: *http://www.checkbiotech.org/*
Roslin Institute: *http://www.ri.bbsrc.ac.uk/*
Science Magazine: *http://www.sciencemag.org/*
The Center for Bioethics and Human Dignity: *http://www.cbhd.org/*
Union of Concerned Scientists: *http://www.ucsusa.org/index.html*

Index

About the Authors

 **Aluízio Borém** is an agronomist with a Ph.D. in molecular genetics from the University of Minnesota. He is currently a faculty member at the Federal University of Vicosa in Brazil and a member of the Brazilian Biosafety Committee, the regulatory agency responsible for biosafety assessment, monitoring, and regulation of biotechnology in Brazil. He is the author of several other books on genetics and crop science.

 **Fabrício R. Santos** is a biologist with a Ph.D. in biochemistry from the Federal University of Minas Gerais (Brazil), where he currently serves as a faculty member. His research interests include the molecular evolution of humans. He is a member of the Brazilian Genome Project and the National Network for DNA Sequencing.

 **David E. Bowen** is an agronomist with an M.S. in applied plant sciences from the University of Minnesota. He is currently pursuing a Ph.D. at the University of Idaho, where he is researching nutritional and genetic improvements in wheat and barley.

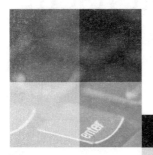

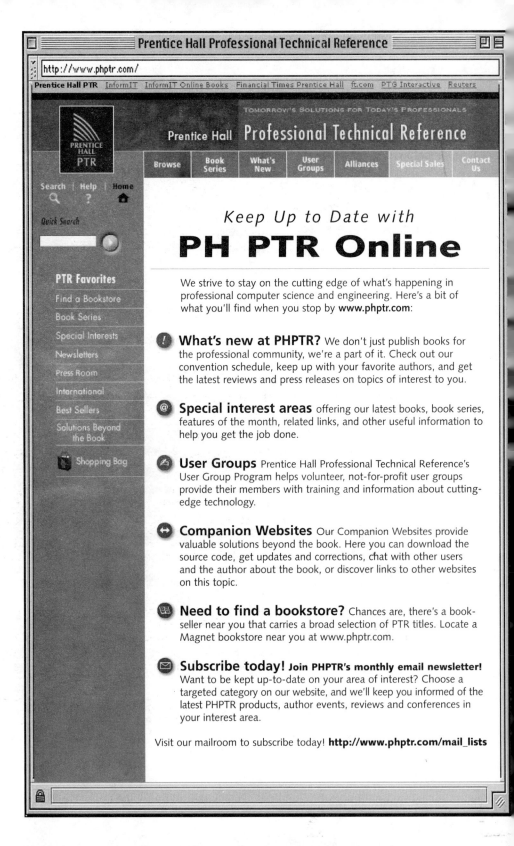